NUCLEAR POWER PLANT SAFETY AND MECHANICAL INTEGRITY

NUCLEAR POWER PLANT SAFETY AND MECHANICAL INTEGRITY

Design and Operability of Mechanical Systems, Equipment and Supporting Structures

GEORGE ANTAKI

and

RAMIZ GILADA

ELSEVIER

Amsterdam • Boston • Heidelberg • London
New York • Oxford • Paris • San Diego
San Francisco • Singapore • Sydney • Tokyo
Butterworth-Heinemann is an imprint of Elsevier

B H

Butterworth-Heinemann is an imprint of Elsevier
The Boulevard, Langford Lane, Kidlington, Oxford, OX5 1GB, UK
225 Wyman Street, Waltham, MA 02451, USA

Notices
Knowledge and best practice in this field are constantly changing. As new research and experience broaden our understanding, changes in research methods, professional practices, or medical treatment may become necessary.

Practitioners and researchers must always rely on their own experience and knowledge in evaluating and using any information, methods, compounds, or experiments described herein. In using such information or methods they should be mindful of their own safety and the safety of others, including parties for whom they have a professional responsibility.

To the fullest extent of the law, neither the Publisher nor the authors, contributors, or editors, assume any liability for any injury and/or damage to persons or property as a matter of products liability, negligence or otherwise, or from any use or operation of any methods, products, instructions, or ideas contained in the material herein.

Library of Congress Cataloging-in-Publication Data
A catalog record for this book is available from the Library of Congress

British Library Cataloguing in Publication Data
A catalogue record for this book is available from the British Library

ISBN: 978-0-12-417248-7

CONTENTS

Biography vii
Preface ix
Acknowledgments xi
Acronyms and Description xiii

1. Regulations, Codes, and Standards **1**
1.1 Requirements 1
1.2 Classification 21

2. Design Basis Loads and Qualification **27**
2.1 Load Definitions 27
2.2 Load Combinations 39
2.3 Dynamic Loads 45
2.4 Vibration Testing and Monitoring 63
2.5 Buried Pipe 73
2.6 Special Analyses 77

3. Examination and Leak Testing **103**
3.1 Examination 103
3.2 Leak Testing 112

4. Suspended Systems **123**
4.1 Piping Systems 123
4.2 Tubing Systems 148
4.3 Heating, Ventilation and Air Conditioning Duct Systems 154
4.4 Conduit Systems 157
4.5 Cable Trays Systems 160

5. Supporting Structures **165**
5.1 Structural Steel 165
5.2 Supporting Concrete 171
5.3 Structural Welding 175
5.4 Bolting to Steel 182
5.5 Bolting to Concrete 183
5.6 Case Studies 187
5.7 Scaffold 200
5.8 Case Study 207
5.9 Component Clearances 209

6. Materials and Aging Mechanisms **213**
6.1 Common Materials 213
6.2 Aging Mechanisms 217

7. Operability and Industry Operating Experience **235**
7.1 Operability 235
7.2 Piping and Tubing Systems 238
7.3 Raceways and Ducts 248
7.4 Mechanical Equipment 252
7.5 Structural 265
7.6 Industry Operating Experience 272
7.7 Examples of IOEs 275

Annex 1: Generic Letters *283*
Annex 2: Bulletins *315*
Index *329*

BIOGRAPHY

Mr Antaki has nearly 40 years of experience in nuclear power and process engineering. He is a Fellow of the American Society of Mechanical Engineers (ASME). He is the Chairman of ASME III Working Group Piping Design and ASME B31.1 Mechanical Design Committee, and a member of the ASME Operation and Maintenance Subgroup. Mr Antaki was an engineer and a group manager at Westinghouse, and is currently Chief Engineer, Becht Nuclear Services. Mr Antaki resides in Aiken, South Carolina.

Mr Gilada has nearly 35 years of experience in nuclear mechanical and civil projects. He is a member of ASME III Working Group Piping Design, the Working Group Flaw Evaluation, and the ASME Operation and Maintenance Subgroup. He is the Principal Consultant Engineer at the Comanche Peak Nuclear Power Plant, Texas, with responsibilities in the field of civil and structural mechanics, and also in piping and suspended systems engineering. Mr Gilada resides in Cleburne, Texas.

PREFACE

This is a book about issues and tasks faced by engineers in nuclear power plants, with a focus on the safe and reliable operation of mechanical systems and equipment, and their supporting structures. The issues and challenges described in this book vary from the most complex subjects, such as replacing steam generators or tackling complex material damage mechanisms, down to the simpler tasks like maintenance, cleaning, and painting that, while simpler, call for logical and technically sound solutions.

This book is a compilation from our personal experiences that span many decades. Our sole purpose is to aid a new generation of engineers navigate through the vast amount of information in this field, and to emphasize the importance, in our business, of approaching every day's challenges with the same degree of rigor and thoroughness to achieve safe and reliable plant operation.

We have tried to describe the thought process followed in defining the issues at hand, and solving them using knowledge from codes, standards, regulations, past experience, and the basic principles of mechanical, materials, and structural engineering. This knowledge converges into a field, best described as Safety and Mechanical Integrity, which we chose for the title. It is the field of engineering in which we face challenges to plant equipment and plant operation (whether human induced, or caused by operational transients, ageing and obsolescence, or natural phenomena hazards) and have to make important operability and run-or-repair decisions.

George Antaki
Becht Engineering Co.

Ramiz Gilada
CPNPP

ACKNOWLEDGMENTS

The opportunity to work in nuclear power plants and to support their safe operation has provided us many opportunities to contribute and also to constantly learn the art of engineering. We are thankful for the support provided to us by plant management in permitting us to share some of this experience for the benefit of others in the industry.

We are honored to be part of Code committees in the field of nuclear power, and thank our colleagues, our friends of so many years on Code committees, for the constant give-and-take that takes place in translating everyday knowledge into codes and standards, and eventually into books like this one.

At times, with the day-to-day emergencies, we thought we may not make our deadlines, but thanks to the excellent support and insights of Chelsea Johnston and Poulouse Joseph of Elsevier, we persevered. Finally, we would like to acknowledge the support we have received from our families during the two year journey to produce this book.

ACRONYMS AND DESCRIPTION

ACI	American Concrete Institute
ACRS	Advisory Committee on Reactor Safeguards
AEC	Atomic Energy Commission
AFW	Auxiliary feedwater
AG	Code on Nuclear Air and Gas Treatment
AHU	Air Handling Unit
AISC	American Institute of Steel Construction
AISI	American Iron and Steel Institute
ANS	American Nuclear Society
ASCE	American Society of Civil Engineers
ASME	American Society of Mechanical Engineers
ASME O&M	Operation and Maintenance of Nuclear Power Plant
ASTM	ASTM International
AWS	American Welding Society
AWWA	American Waterworks Association
BDBE	Beyond-Design Basis Event (and BDB Earthquake)
BEZ	Break Exclusion Zone
BL	NRC Bulletin
BTP	Branch technical position
BWR	Boiling water reactor
B&PV	Boiler and Pressure Vessel
CC	Code Case
CEA-CEN	Commissariat a l'Energie Atomique Centre d'Etudes Nucleaires
CFD	Computational fluid dynamics
CFR	Code of Federal Regulations
CGD	Commercial Grade Dedication
CP	Construction permit
CR	Contractor report
CRDM	Control rod drive mechanism
CS	Core support and Containment spray
DMW	Dissimilar metal welds
EPRI	Electric Power Research Institute
EQ	Environmental qualification and Equipment qualification
ERDA	Energy Research and Development Administration
ET	Eddy current testing
FA	Functionality assessment
FAC	Flow-accelerated corrosion
FEA	Finite element analysis
FEMA	Federal Emergency Management Agency
FIV	Flow-induced vibration
FLEX	Diverse and flexible coping strategies

FMEA	Failure mode and effects analysis
FSAR	Final safety analysis report
GDC	General design criteria
GIP	Generic implementation procedure
GL	NRC Generic Letter
GMRS	Ground motion response spectra
GPM	Gallons per minute
HE	High Energy
HVAC	Heating, ventilation and air conditioning
IAEA	International Atomic Energy Agency
ID	Inside diameter
IEEE	Institute of Electrical and Electronics Engineers
IER	Industry Event Report
IN	NRC Information Notice
INPO	Institute of Nuclear Power Operations
IOE	Industry operating experience
ISI	Inservice inspection
IST	Inservice testing, ISTB (pumps), ISTC (valves), ISTD (snubbers)
LBB	Leak before break
LER	Licensee Event Report
LOCA	Loss of coolant accident
MC	Metal containment
ME	Moderate energy
MEB	Mechanical engineering branch
MR	Maintenance rule
MRP	Material Reliability Program (EPRI)
MSIV	Main steam isolation valve
MSS-SP	Manufacturers Standardization Society – Standard Practices
MT	Magnetic testing
NDE	Nondestructive examination
NDT	Nondestructive testing
NEI	Nuclear Energy Institute
NEMA	National Electrical Manufacturers Association
NQA	Nuclear Quality Assurance
NRC	Nuclear Regulatory Commission
NUREG	NRC Regulatory Guidance
OBE	Operating basis earthquake
OD	Operability determination
OD	Outside diameter
OSHA	Occupational Safety and Health Administration
O&M	Operation and Maintenance
PCC	Post-Construction Code
PRA	Probabilistic risk assessment

PRHA	Pipe rupture hazards analysis
PT	Penetrant testing
PTS	Pressurized thermal shock
PVRC	Pressure Vessel Research Council
PWR	Pressurized water reactor (and pipe whip restraint)
PWSCC	Primary water stress corrosion cracking
P&ID	Piping and instrumentation diagram
QA	Quality assurance
QC	Quality control
QME	Qualification of Active Mechanical Equipment
RAHA	Radial Arm and Hoist Assembly
RCS	Reactor coolant system
RG	Regulatory guide
RHR	Residual heat removal
RT	Radiographic testing
SAM	Seismic anchor motion
SAR	Safety analysis report
SCC	Stress corrosion cracking
SG	Steam generator
SMA	Seismic margins assessment
SMACNA	Sheet Metal & Air Conditioning Contractors' National Association
SPRA	Seismic Probabilistic Risk Assessment
SQUG	Seismic Qualification Utilities Group
SRP	Standard review plan
SS	Stainless steel
SSC	Structure, system and component
SSE	Safe shutdown earthquake
TIG	Tungsten inert gas (welding)
TS	Technical specification
UT	Ultrasonic testing
VT	Visual testing
ZOI	Zone of influence

Section III

NCA	ASME Section III Subsection General Requirements for Division 1 (Piping & Components) and Division 2 (Concrete Containments)
NB	ASME III Subsection for Class 1 Components
NC	ASME III Subsection for Class 2 Components
ND	ASME III Subsection for Class 3 Components
NE	ASME III Subsection for Class MC Components
NF	ASME III Subsection for Pipe Supports

Section XI

IWA	ASME XI General Requirements
IWB	ASME XI Requirements for Class 1 Components of Light-Water Cooled Plants
IWC	ASME XI Requirements for Class 2 Components of Light-Water Cooled Plants
IWD	ASME XI Requirements for Class 3 Components of Light-Water Cooled Plants

CHAPTER 1

Regulations, Codes, and Standards

Chapter 1 covers the regulations that govern the design, construction, and operation of mechanical systems and components in nuclear power plants. Regulations are the building blocks of nuclear power plant engineering; they are defined in each plant's safety analysis report (SAR), and they must be understood and followed at each step of the engineering process. Following the regulations, we address Codes and Standards, with emphasis on the American Society of Mechanical Engineers (ASME) Boiler and Pressure Vessel Code (B&PV Code) because it covers piping systems, vessels, pumps, and valves, which are of particular interest in this book. We close this chapter explaining how structures, systems, and components (SSCs) are classified into safety classes, seismic categories, and essential classes.

1.1 REQUIREMENTS

1.1.1 Regulation

Why start the book with regulations?

Engineering activities at a nuclear power plant are highly regulated. An engineer in a nuclear power plant has to focus on three types of requirements, all of them essential and complementary: (1) following the company procedures, (2) following good engineering practice (codes, standards, guides, etc.), and (3) Following the regulations applicable at the plant.

It is useful to first look at the regulatory aspect of nuclear plant engineering in the United States, before turning our attention to codes and standards, because nuclear power regulations control the application of codes and standards, and provide supplementary requirements that are not contained in codes and standards. Also, regulations control how to document engineering issues and their resolution.

How are engineering activities regulated in the nuclear power industry?

The engineering of nuclear power plants is closely regulated by the US Nuclear Regulatory Commission (NRC). The NRC was formed from the

Nuclear Power Plant Safety and Mechanical Integrity
ISBN 978-0-12-417248-7

Atomic Energy Commission (AEC) through the Energy Reorganization Act of 1974, which replaced the Atomic Energy Act of 1954. It made the NRC a regulatory agency, no longer involved in the development and promotion of nuclear power as was its predecessor the AEC.

The NRC is empowered by the Congress to regulate and provide oversight of the complete life cycle of nuclear power plants, from design and construction, through operation, life extension, and decommissioning.

What are the functions of the NRC?

The NRC has several functions, which include: (1) to develop regulations and guidance, (2) to license applicants to operate the nuclear facility; (3) to oversee compliance with the licensing basis and safety requirements of the plant; (4) to evaluate operational experience and communicate lessons learned, and to impose plant-specific or industry-wide actions if necessary; (5) to conduct research; and (6) to hold hearings to address safety concerns related to nuclear plant operations.

How is the NRC organized?

At the top of the NRC structure are five NRC Commissioners, including the chairperson. Reporting to the Commissioners is the NRC staff and the Advisory Committee on Reactor Safeguards (ACRS). Within the NRC staff, under the Executive Director of Operations, there are several departments, including four geographical regions, the office of Nuclear Reactor Regulation, the office of New Reactors, and the office of Nuclear Regulatory Research.

What is the ACRS?

The ACRS is an advisory group of foremost experts, independent of the NRC staff, who report directly to the Commission, with four primary objectives: (1) to review safety studies, (2) to advise the Commission on safety standards, (3) to review generic safety topics, and (4) to advise on radiation protection.

How are the regulations organized?

At the top of the hierarchy of regulations is the Code of Federal Regulations (CFR), for nuclear power the CFRs of interest are title 10, referred to as 10 CFR, in particular:

- 10 CFR Part 20 "Standards for Protection Against Radiation"
- 10 CFR Part 21 "Reporting of Defects and Noncompliance"

- 10 CFR Part 50 "Domestic Licensing of Production and Utilization Facilities"
- 10 CFR Part 100 "Reactor Site Criteria"

To what extent do the CFRs affect engineering?

The CFRs are the foundations over which the plant's SAR, codes, standards, and procedures are developed. They permeate every aspect of engineering. A good example is 10 CFR section 50.72 "Immediate Notification Requirements for Operating Nuclear Power Reactors." This CFR section has specific requirements regarding reporting of conditions and activities to the NRC Operations Center. Certain conditions must be reported in as little as 4 hours.

What are the upper-level safety requirements?

10 CFR Part 50 contains Appendix A, which spells out upper-level general design criteria (GDC). The following GDCs are of particular interest in the design and qualification of SSCs:

- Criterion 1 "Quality Standards and Records"
- Criterion 2 "Design Bases for Protection Against Natural Phenomena"
- Criterion 3 "Fire Protection"
- Criterion 4 "Environmental and Missile Design Bases"
- Criterion 5 "Sharing of Structures, Systems, and Components"
- Criterion 13 "Instrumentation and Controls"
- Criterion 14 "Reactor Coolant Pressure Boundary"
- Criterion 15 "Reactor Coolant System Design"
- Criterion 16 "Containment Design"
- Criterion 17 "Electric Power Systems"
- Criterion 18 "Inspection and Testing of Electric Power Systems"
- Criterion 19 "Control Room"
- Criterion 21 "Protection System Reliability and Testability"
- Criterion 22 "Protection System Independence"
- Criterion 23 "Protection System Failure Modes"
- Criterion 24 "Separation of Protection and Control Systems"
- Criterion 30 "Quality of Reactor Coolant Pressure Boundary"
- Criterion 34 "Residual Heat Removal"
- Criterion 35 "Emergency Core Cooling"
- Criterion 38 "Containment Heat Removal"
- Criterion 41 "Containment Atmosphere Cleanup"
- Criterion 44 "Cooling Water"

- Criterion 50 "Containment Design Basis"
- Criterion 53 "Provisions for Containment Testing and Inspection"
- Criterion 54 "Piping Systems Penetrating Containment"
- Criterion 55 "Reactor Coolant Penetrating Containment"
- Criterion 56 "Primary Containment Isolation"
- Criterion 57 "Closed System Isolation Valves"
- Criterion 60 "Control of Releases of Radioactive Materials"
- Criterion 61 "Fuel Storage and Handling and Radioactivity Control"

Other 10 CFR 50 appendices of importance to the subject of this book are Appendix B "Quality Assurance Criteria for Nuclear Power Plants and Fuel Reprocessing Plants"; Appendix J "Primary Reactor Containment Leakage Testing for Water-Cooled Power Reactors"; and Appendix R "Fire Protection Program for Nuclear Power Facilities."

What is the next level of regulation, below the CFRs?

At the next engineering level is the Standard Review Plan (SRP) which details the engineering methods and criteria to be followed in designing, qualifying, and operating the plant. Each plant commits to meet the SRP and documents its compliance in the SAR. Plants have a Preliminary Safety Analysis Report at the construction permit stage, and a Final Safety Analysis Report (FSAR) at the operating license stage. The commitments made in the plant FSAR will guide the engineering activities throughout the plant lifetime. New plants have a preapproved design basis, captured in a Design Control Document, which also follows the SRP.

The SRP is (US Nuclear Regulatory Commission Regulation) NUREG-0800 "Standard Review Plan for the Review of Safety Analysis Reports for Nuclear Power Plants." Until 1975 it was published as Regulatory Guide (RG) 1.70 "Standard Format and Content of Safety Analysis Reports for Nuclear Power Plants." The SRP describes the methods and criteria that are acceptable for the design, construction, operation, inspection, and repair of nuclear power plants. Regarding SSCs, the SRP is in many ways a technical guide to preparing project-specific engineering design, analysis, and qualification procedures. The SRP is subdivided into 19 chapters, and each chapter addresses a specific engineering discipline. The 19 chapters are:

- Chapter 1: Introduction and Interfaces
- Chapter 2: Site Characteristics and Site Parameters
- Chapter 3: Design of Structures, Components, Equipment, and Systems

- Chapter 4: Reactor
- Chapter 5: Reactor Coolant System and Connected Systems
- Chapter 6: Engineered Safety Features
- Chapter 7: Instrumentation and Controls
- Chapter 8: Electric Power
- Chapter 9: Auxiliary Systems
- Chapter 10: Steam and Power Conversion System
- Chapter 11: Radioactive Waste Management
- Chapter 12: Radiation Protection
- Chapter 13: Conduct of Operations
- Chapter 14: Initial Test Program and ITAAC-Design Certification
- Chapter 15: Transient and Accident Analysis
- Chapter 16: Technical Specifications
- Chapter 17: Quality Assurance
- Chapter 18: Human Factors Engineering
- Chapter 19: Severe Accidents

Which chapter addresses design and integrity of mechanical systems and components?

Of particular interest to us is Chapter 3 of the SRP which addresses the design and qualification of SSCs. Chapter 3 contains several sections, listed here. The first version of Chapter 3 dates back to 1975, except as noted in parenthesis. There used to be a Section 3.1 which addressed compliance with the GDCs of 10 CFR 50 Appendix A.

- 3.2.1: Seismic Classification
- 3.2.2: System Quality Group Classification
- 3.3.1: Wind Loadings
- 3.3.2: Tornado Loadings
- 3.4.1: Internal Flood Protection for Onsite Equipment Failures
- 3.4.2: Analysis Procedures
- 3.5.1.1: Internally Generated Missiles (Outside Containment)
- 3.5.1.2: Internally Generated Missiles (Inside Containment)
- 3.5.1.3: Turbine Missiles
- 3.5.1.4: Missiles Generated by Tornadoes and Extreme Winds
- 3.5.1.5: Site Proximity Missiles (Except Aircraft)
- 3.5.1.6: Aircraft Hazards
- 3.5.2: Structures, Systems, and Components to be Protected from Externally Generated Missiles
- 3.5.3: Barrier Design Procedures

- 3.6.1: Plant Design for Protection Against Postulated Piping Failures in Fluid Systems Outside Containment
- 3.6.2: Determination of Rupture Locations and Dynamic Effects Associated with the Postulated Rupture of Piping
- 3.6.3: Leak-Before-Break Evaluation Procedures
- BTP 3-3: Protection Against Postulated Piping Failures in Fluid Systems Outside Containment
- BTP 3-4: Postulated Rupture Locations in Fluid System Piping Inside and Outside Containment
- 3.7.1: Seismic Design Parameters
- 3.7.2: Seismic System Analysis
- 3.7.3: Seismic Subsystem Analysis
- 3.7.4: Seismic Instrumentation
- 3.8.1: Concrete Containment
- 3.8.2: Steel Containment
- 3.8.3: Concrete and Steel Internal Structures of Steel or Concrete Containments
- 3.8.4: Other Seismic Category I Structures
- 3.8.5: Foundations
- 3.9.1: Special Topics for Mechanical Components
- 3.9.2: Dynamic Testing and Analysis of Systems, Structures, and Components
- 3.9.3: ASME Code Class 1, 2, and 3 Components, and Component Supports, and Core Support Structures
- 3.9.4: Control Rod Drive Systems
- 3.9.5: Reactor Pressure Vessel Internals
- 3.9.6: Functional Design, Qualification, and Inservice Testing Programs for Pumps, Valves, and Dynamic Restraints
- 3.9.7: Risk-Informed Inservice Testing (1998)
- 3.9.8: Risk-Informed Inservice Inspection of Piping (2008)
- 3.10: Seismic and Dynamic Qualification of Mechanical and Electrical Equipment
- 3.11: Environmental Qualification of Mechanical and Electrical Equipment
- 3.12: ASME Code Class 1, 2, and 3 Piping Systems, Piping Components, and their Associated Supports (2007)
- 3.13: Threaded Fasteners - ASME Code Class 1, 2, and 3 (1996)

What are RGs?

RGs are engineering guidance documents issued by the NRC which address a specific technical topic. Many RGs are called out in the SRP. Examples of RGs of interest to the design and qualification of SSCs include:

Quality Assurance RGs:

- RG 1.26 "Quality Group Classification and Standards for Water-, Steam-, and Radioactive-Waste-Containing Components of Nuclear Power Plants"
- RG 1.28 "Quality Assurance Program Requirements (Design and Construction)"
- RG 1.33 "Quality Assurance Program (Operation)"
- RG 1.37 "Quality Assurance Requirements for Cleaning of Fluid Systems and Associated Components of Water-Cooled Nuclear Power Plants"
- RG 1.54 "Quality Assurance Requirements for Protective Coatings Applied to Water-Cooled Nuclear Power Plants"
- RG 1.64 "Quality Assurance Requirements for the Design of Nuclear Power Plants"
- RG 1.74 "Quality Assurance Terms and Definitions"
- RG 1.88 "Collection, Storage, and Maintenance of Nuclear Power Plant Quality Assurance Records"
- RG 1.116 "Quality Assurance Requirements for Installation, Inspection, and Testing of Mechanical Equipment and Systems"
- RG 1.123 "Quality Assurance Requirements for Control of Procurement of Items and Services for Nuclear Power Plants"

Loads and Mechanical and Structural Design RGs:

- RG 1.29 "Seismic Design Classification"
- RG 1.46 "Protection Against Pipe Whip Inside Containment"
- RG 1.48 "Design Limits and Loading Combinations for Seismic Category I Fluid System Components"
- RG 1.57 "Design Limits and Loading Combinations for Metal Primary Reactor Containment System Components"
- RG 1.60 "Design Response Spectra for Seismic Design of Nuclear Power Plants"
- RG 1.61 "Damping Values for Seismic Design of Nuclear Power Plants"
- RG 1.76 "Design Basis Tornado for Nuclear Power Plants"

- RG 1.92 "Combining Modal Responses and Spatial Components in Seismic Response Analysis"
- RG 1.100 "Seismic Qualification of Electric Equipment for Nuclear Power Plants"
- RG 1.115 "Protection Against Low-Trajectory Turbine Missiles"
- RG 1.117 "Tornado Design Classification"
- RG 1.122 "Development of Floor Design Response Spectra for Seismic Design of Floor-Supported Equipment or Components"
- RG 1.124 "Design Limits and Loading Combinations for Class 1 Linear-Type Component Supports"
- RG 1.130 "Service limits and Loading Combinations for Class 1 Plate- and shell-Type Component Supports"
- RG 1.143 "Design Guidance for Radioactive Waste Management Systems, Structures, and Components Installed in Light-Water-Cooled Nuclear Power Plants"
- RG 1.207, "Guidelines for Evaluating Fatigue Analyses Incorporating the Life Reduction of Metal Components Due to the Effects of the Light-Water Reactor Environment for New Reactors"

Materials RGs:

- RG 1.36 "Nonmetallic Thermal Insulation for Austenitic Stainless Steel"
- RG 1.44 "Control of the Use of Sensitized Stainless Steel"

Performance and Testing RGs:

- RG 1.45 "Reactor Coolant Pressure Boundary Leakage Detection Systems"
- RG 1.53 "Application of the Single-Failure Criterion to Nuclear Power Plant Protection Systems"
- RG 1.68 "Initial Test Programs for Water-Cooled Nuclear Power Plants"
- RG 1.73 "Qualification Tests of Electric Valve Operators Installed Inside the Containment of Nuclear Power Plants"
- RG 1.89 "Qualification of Class 1E Equipment for Nuclear Power Plants"
- RG 1.131 "Qualification Tests of Electric Cables, Field Splices, and Connections for Light-Water-Cooled Nuclear Power Plants"

What are NUREGs?

NUREGs are described in the NRC Web site as "*Reports or brochures on regulatory decisions, results of research, results of incident investigations, and other technical and administrative information.*" When prepared by the NRC staff they are labeled NUREG, and when prepared by a contractor (oftentimes a National Laboratory) they are labeled NUREG/contractor report (CR). They provide a wealth of technical knowledge, some in the form of research results, incident reports, and studies. In many cases NUREGs provide the technical backbone to regulations. They have also guided national codes and standards in developing or improving the design, fabrication, and inspection rules of the codes.

So many NUREGs are of interest to the design and integrity of nuclear systems and components that is not possible to list them all, or even to compile a nearly complete list. All we can do is list as a sample of NUREGs of interest to mechanical distribution systems:

- NUREG/CR-0371 "Stress Indices for Girth Welded Joints Including Radial Weld Shrinkage, Mismatch, and Tapered-Wall Transitions"
- NUREG-0484 "Methodology for Combining Dynamic Responses"
- NUREG-0554 "Single-Failure-Proof Cranes for Nuclear Power Plants"
- NUREG-0582 "Water Hammer in Nuclear Power Plants"
- NUREG-0612 "Control of Heavy Loads at Nuclear Power Plants"
- NUREG-0737 "Clarification of TMI Action Plan Requirements"
- NUREG/CR-0778 "Stress Indices and Flexibility Factors for Nozzles in Pressure Vessel and Piping"
- NUREG-1030 "Seismic Qualification of Equipment in Operating Nuclear Power Plant"
- NUREG-1061 "Report of the US Nuclear Regulatory Commission Piping Review Committee"
- NUREG/CR-1161 "Recommended Revision to Nuclear Regulatory Commission Seismic Design Criteria"
- NUREG/CR-1980, BLN-NUREG-51357 "Dynamic Analysis of Piping Using the Structural Overlap Method"
- NUREG/CR-2913 "Two Phase Jet Loads"

What are generic letters, bulletins, and information notices?

These are means for the NRC to communicate to the industry generic information (Information Notices (IN)), or generic actions (Bulletins and Generic Letters (GLs)) which affect multiple plants. The first digits in an IN or a GL are the year the document was first issued. A list of, Bulletins, and GLs are provided in Annex.

1.1.2 Codes and Standards

Which codes apply to mechanical systems at a nuclear power plant?

Safety-related SSCs are designed, fabricated in the shop, erected in the field, examined, tested, and inspected in service in accordance with Codes and Standards. These Codes and Standards, including the edition and addenda, are only applicable to the extent that they have been approved by the NRC for use at a particular plant, as documented in the plant FSAR.

The Codes for nuclear power plant mechanical systems include the ASME B&PV Code for vessels, piping, pumps, and valves; the ASME AG-1 Code (Air and Gas) for cooling and ventilation systems; the ASME B31.1 Code for power piping; the ASME Operation and Maintenance (O&M) Code for the operation and maintenance of active mechanical equipment and components; the ASME Qualification of Mechanical Equipment (QME) Code; the American Institute of Steel Construction (AISC) Code for design of certain steel structures; the American Iron and Steel Institute (AISI) code for design of some support members; and the American Concrete Institute (ACI) Code for the design of concrete structures.

Through which regulation is the ASME B&PV code applicable for nuclear power plants?

The ASME B&PV Code is called out explicitly in 10 CFR 50.55a, which states, in section (b)(2): "Systems and components of boiling and pressurized water-cooled nuclear power reactors must meet the requirements of the ASME Boiler and Pressure Vessel Code specified in paragraphs (b), (c), (d), (e), (f), and (g) of this section. Protection systems of nuclear power reactors of all types must meet the requirements specified in paragraph (h) of this section."

But not all code editions are approved, and the NRC may take exceptions to the provisions of a particular Code edition or addendum, which will be spelled out in 10 CFR 50.55(a). A specific ASME III and ASME XI Code edition and addendum of the ASME B&PV Code, and specific Code Cases (CCs) are approved at each individual nuclear power plant.

What should we know about the origin of the ASME B&PV code?

As an engineer, it is important to not use Codes blindly, but to understand the technical basis of the Code requirements, and even the origin, history, and evolution of these requirements.

The ASME B&PV codes had their origin in 1880, when a group of industrialists and technical professionals gathered in New York to create standards for the manufacturing of tools and machine parts. The ASME was founded from this gathering in New York City.

The ASME B&PV Code received its impetus from the steam-powered industrial revolution, with steam boilers taking centerstage. But in their infancy, boilers were not reliable; they needed frequent maintenance, and in some cases they would malfunction and even explode. In 1905, a vessel explosion in a shoe factory in Brockton, MA, USA, caused many fatalities and much damage to several building blocks. This accident caused Massachusetts to establish a Board of Boiler Rules, which published its state law in 1908. As other cities and states followed suit, there was a concern that a boiler fabricated in one state could not be sold in another: There was a need for uniformity. A Boiler Code Committee was formed within ASME, in 1911, and the first edition of the ASME B&PV Code, was published in 1915. The B&PV Code was later incorporated into state laws.

Over time, the ASME B&PV Code expanded its scope to cover not only boilers, but all sorts of pressure vessels, their materials, design, fabrication, inspection, and testing.

What B&PV code sections apply today to nuclear power plants?

The ASME B&PV Code is composed of several "Sections" that address the design of B&PVs and associated systems and components. For nuclear power plants, the ASME Codes of interest are:

- Section II for materials
- Section III for design and construction
- Section V for examination and leak testing techniques
- Section IX for welding
- Section XI for in-service inspection
- NQA-1 (National Quality Assurance) for quality assurance
- O&M for operations and maintenance of mechanical equipment
- QME for the qualification of mechanical equipment
- AG-1 for air handling systems

What is the scope and structure of section III of the ASME B&PV code?

Section III of the ASME B&PV Code consists of five Divisions, they are:

- Division 1: Nuclear Components
- Division 2: Concrete Containments
- Division 3: Containments for Transport and Storage of Nuclear Fuel and High Level Radioactive Material and Waste
- Division 4: Fusion Reactors
- Division 5: Elevated Temperature

From here onward, we will refer to ASME B&PV Code Section III Division 1 simply as ASME III.

Does ASME III apply in all US nuclear power plants?

No. The older plants were licensed and designed before the advent of the ASME III nuclear code. Up to 1963, nuclear power plants were designed to ASME VIII, with supplementary requirements spelled out in their FSAR. In 1963, ASME III was published for the design and fabrication of safety-related nuclear vessels. The number, Section III, was reassigned from the then defunct ASME III Code for locomotives. Safety-related piping systems were still designed and fabricated to ASME B31.1, with supplementary requirements, in the form of FSAR and nuclear piping CCs, the first one being CC N-1 in 1962. In 1971 ASME III was expanded to cover piping, pumps, and valves, and this remains the scope of ASME III to this day: pressure vessels, pumps, valves, piping, and tanks.

Every 3 years, a new edition was issued, and addenda were published within a year. Plants were therefore designed to different editions and addenda, depending on the date of their construction license. Starting in 2011, editions are being published every 2 years, without addenda.

What is the structure of ASME III division 1?

First are the general requirements of NCA. This is an important Section of ASME Section III titled "General Requirements for Division 1 (Piping & Components) and Division 2 (Concrete Containments)". NCA is composed of the following subsections:

- NCA-1000: Scope of Section III
- NCA-2000: Classification of Components and Supports
- NCA-3000: Responsibilities and Duties
- NCA-4000: Quality Assurance
- NCA-5000: Authorized Inspection
- NCA-7000: Reference Standards
- NCA-8000: Certificates, Nameplates, Code Symbol Stamping, and Data Reports
- NCA-9000: Glossary

NCA is followed by several subsections that apply to various classes of components:

- Class 1: Subsection NB (vessels, piping, pumps, and valves)
- Class 2: Subsection NC (vessels, piping, pumps, and valves)
- Class 3: Subsection ND (vessels, piping, pumps, and valves)

- Class MC: Subsection NE (metal containment)
- Class CS: Subsection NG (core support)
- Supports: Subsection NF (component supports)
- Internal structures: Subsection NG

Class 1 SSCs (Subsection NB) are applicable to the reactor vessel and piping systems and components directly connected to them. Class 1 (NB) requirements are more stringent than Class 2 and 3 (NC and ND) requirements.

How are ASME III division 1 NB/NC/ND organized?

The structure of NB, NC, and ND is uniform among Classes 1, 2, and 3, as follows:

- 1000: Introduction
- 2000: Materials
- 3000: Design
- 4000: Fabrication and Installation
- 5000: Examination
- 6000: Testing
- 7000: Overpressure Protection
- 8000: Nameplates, Stamps, Reports

Finally, the structure of 3000 itself is also uniform among Classes 1, 2, and 3, as follows:

- NB/NC/ND-3100 General Design
- NB/NC/ND-3200 Design by Analysis
- NB/NC/ND-3300 Vessel Design
- NB/NC/ND-3400 Pumps
- NB/NC/ND-3500 Valves
- NB/NC/ND-3600 Piping
- NC/ND-3800 Storage tanks (atm)
- NC/ND-3900 Storage tanks (atm to 15 psig)

Are there means, other than editions or addenda, by which ASME III is kept up-to-date?

When a new matter, not addressed in the current ASME III, or when an alternative means of addressing a matter is developed, it is often published as an ASME III Code Case (CC).

Which ASME CCs are commonly used for design and qualification of SSCs in nuclear power plants?

CCs are published by the ASME, and can only be used in a nuclear power plant if they are explicitly approved for use, and if they are used within the conditions of applicability of the CC, published by the NRC in the following RGs:

- RG 1.84 "Design and Fabrication Code Case Acceptability ASME Section III Division I"
- RG 1.85 "Materials Code Case Acceptability ASME III Division I"
- RG 1.147 "Inservice Inspection Code Case Acceptability ASME XI, Division 1"
- RG 1.193 "ASME Code Cases not Approved for Use"

Examples of nuclear CCs that are commonly used for safety-related SSCs, within the limits specified in the above RGs include:

- N-62, Internal and External Valve Items, Section III, Division 1, Classes 1, 2, and 3
- N-71, Additional Material for Component Supports, Section III, Division 1, Subsection NF, Classes 1, 2, and 3, and MC Component supports.
- N-108, Deviation from the specified silicon ranges in ASME Material Specifications, Section III, Division 1 and VIII Divisions 1 and 2.
- N-180, Examination of springs for Class 1 component standard supports
- N-224, Use of American Society of Testing and Materials (ASTM) A500 Grade B and ASTM A501 Structural Tubing for Welded Attachments for Section III, Class 2 and 3 & MC Construction.
- N-225, Certification and Identification of Material for Component Supports, Section III, Division 1
- N-242, Material Certification, Section III, Division 1, Classes 1, 2, 3 MC (Metal Containment) and CS (Core Support) construction.
- N-247, Certified Design Report Summary for Component Standard Supports, Section III, Division 1, Class 1, 2, 3, and MC
- N-249, Additional Materials for Component Supports Fabricated without Welding, Section III, Division 1, Subsection NF, Class 1, 2, 3 and MC Component Supports
- N-249, Additional Materials for Subsection NF Class 1, 2, and 3, and MC Component Supports Fabricated Without Welding Section III, Division 1.
- N-253, Construction of Class 2 or Class 3 components for Elevated Temperature Service.
- N-263, dated March 17, 1980, Alternate Thread Forms, Series, and Fits for Component Supports Section III, Division 1.
- N-309, Identification of Materials for Component Supports, Section III, Division 1.
- N-310, Certification of Bolting Materials.

- N-318, Procedure for Evaluation of the Design of Rectangular Cross Section Attachment on Class 2 or 3 Piping, Section III, Division 1. Any design satisfying Revision 1 of this CC can automatically be considered satisfying Revision 2.
- N-392, Procedure for Evaluation of the Design of Hollow Circular Cross Section Welded Attachments on Classes 2 and 3 Piping, Section III, Division 1
- N-397, Alternative Rules to the Spectral Broadening Procedures of N-1226.3 for Class 1, 2, and 3 Piping, Section III, Division 1.
- N-411, Alternative Damping Values for Seismic Analysis of Class 1, 2, and 3 Piping Systems, Section III, Division 1.
- N-413, Minimum Size of Fillet Welds for Subsection NF Linear-Type Supports, Section III, Division 1, Class 1, 2, and 3
- N-420, Linear Energy Absorbing Supports for Subsection NF, Classes 1,2 and three Construction, Section III, Division 1
- N-513, Evaluation Criteria for Temporary Acceptance of Flaws in Moderate Energy Class 2 or 3 Piping, Section XI, Division 1
- N-597, Requirements for Analytical Evaluation of Pipe Wall Thinning Section XI, Division 1
- N-1606, Stress Criteria Section III, Class 2 and 3 Piping Subject to Upset, Emergency, and Faulted Operating Conditions
- N-1636, Stress Criteria for Section III, Class 2 and 3 Pumps Subjected to Upset, Emergency, and Faulted Operating Conditions
- N-1644, Additional Materials for Component Supports, and Alternate Design Requirements for Bolted Joints, Subsection NF, Class 1,2,3, and MC Construction
- N-1683, Bolt Holes for Section III, Class 1, 2, 3 and MC Component Supports
- N-1685, Furnace Brazing Section III, Class 1, 2, 3 and MC Construction
- N-1686, Furnace Brazing, Section III, Subsection NF, Component Supports
- N-1706, Data Report Forms for Component Supports, Section III, Class 1, 2, 3
- N-1724, Deviation from the Specified Silicon Ranges in ASME Material Specifications, Section III, Division 1 and Section VIII, Division 1 and 2.
- N-1728, Steel Structural Shapes and Small Material Products for Component Supports, Section III, Division I, Construction.

- N-1729, Minimum Edged Distance - Bolting for Section III Division 1, Class 1, 2, 3 and MC Construction of Component Supports
- N-1734, Weld Design for Use for Section III, Division 1, Class 1, 2, 3 and MC Construction of component supports.
- N-1745, Stress Indices for Structural Attachments, Class 1, Section III, Divisions 2

What other codes, besides ASME III, apply to nuclear power plants?

While ASME III covers the design and fabrication of nuclear plant vessels, pumps, valves, piping, and tanks, and their supports, other codes address other aspects of the plant, depending on the type of commodity (structural, mechanical, electrical, etc.) and the function (design, manufacturing, operation, inspection, etc.). The following codes and standards are applicable to mechanical systems in the nuclear industry:

The ASME B&PV Codes other than ASME III are:

- ASME II for materials
- ASME V for examination and testing
- ASME IX for welding
- ASME XI for in-service inspections and repairs

The Codes for other commodities are:

- American Institute of Steel Construction (AISC)
- American Iron and Steel Institute (AISI)
- American Concrete Institute (ACI)
- American National Standard Institute (ANSI)
- American Nuclear Society (ANS)
- ASTM International (ASTM)
- American Welding Society (AWS)
- Institute of Electrical and Electronics Engineers (IEEE)
- National Fire Protection Association
- Underwriters Laboratory
- International Electrotechnical Commission
- American Water Works Association

What is the scope of ASME XI?

Section XI addresses in-service inspection (ISI), and the characterization and evaluation of flaws detected during ISIs. It also addresses repairs and replacements of static (fixed) equipment, vessels, and piping systems, and the pressure boundary of pumps and valves. In-service testing of active components (pumps, valves, snubbers) is not addressed in ASME XI; it is addressed in ASME O&M.

"There are parallel in-service inspection codes developed by the non-nuclear industries, and which provide useful practical guidance. They include API 510 and 570 for the inspection of vessels and piping, API 579-1/ASME FFS-1 for fitness-for-service of pressure equipment, ASME Post-Construction Codes PCC-1 for bolted joints, PCC-2 for repairs, and PCC-3 for risk-based inspections."

What is the scope and the structure of ASME O&M?

Up to 2009, what is now the single ASME Code for Operation and Maintenance of Nuclear Power Plants (ASME O&M) consisted of two separate publications: The O&M Code and the O&M Standards and Guides. Today, all these parts are combined into a single ASME O&M Code which consists of the following three Divisions:

Division 1, Section Inservice Testing (IST) "Light-Water Reactor Nuclear Power Plants" contains the requirements for preservice and in-service testing and examination of components to assess their operational readiness in light water nuclear power plants, Section IST includes:

- Subsection ISTA "General Requirements"
- Subsection ISTB "Inservice Testing of Pumps"
- Subsection ISTC "Inservice Testing of Valves"
- Subsection ISTD "Preservice and Inservice Examination and Testing of Dynamic Restraints (Snubbers)"
- Subsection ISTE "Risk-Informed Inservice Testing of Components"

Division 2, Standards for "Light-Water Reactor Power Plants" addresses the following topics:

- Part 3 "Vibration Testing of Piping Systems"
- Part 12 "Loose Part Monitoring"
- Part 16 "Performance Testing and Inspection of Diesel Drive Assemblies"
- Part 21 "Inservice Performance Testing of Heat Exchangers"
- Part 24 "Reactor Coolant and Recirculation Pump Condition Monitoring"
- Part 26 "Determination of Reactor Coolant Temperature from Diverse Measurements"
- Part 28 "Standard for Performance Testing of Systems"
- Part 29 "Alternative Treatment Requirements for RISC-3 Pumps and Valves"

Division 3, Guides "Light-Water Reactor Power Plants" addresses the following topics:

- Part 5 "Inservice Monitoring of Core Support Barrel Axial Preload in Pressurized Water Reactor Power Plants"
- Part 7 "Requirements for Thermal Expansion Testing of Nuclear Power Plant Piping Systems"
- Part 11 "Vibration Testing and Assessment of Heat Exchangers"
- Part 14 "Vibration Monitoring of Rotating Equipment in Nuclear Power Plants"
- Part 19 "Preservice and Periodic Performance Testing of Pneumatically and Hydraulically Operated Valve Assemblies"
- Part 23 "Inservice Monitoring of Reactor Internals Vibration in Pressurized Water Reactor Power Plants"

How do the AISC and AISI codes apply?

The application of the AISC "Manual of Steel Construction" and AISI "Cold-Formed Steel Design Manual" codes depends on the vintage of the plant. For the older plants, steel structures were designed to the AISC and AISI. For the more recent plants, support structures for safety class systems and components are designed and constructed to ASME III NF. AISC N-690 was used for special structures such as whip restraints.

How does the ACI code apply?

The ACI codes address concrete construction, both reinforced and unreinforced structures, as well as embedments in concrete and cast-in-place and postinstalled anchor bolts. Two important ACI codes used in the nuclear power industry are:

- ACI-318 "Building Code Requirements for Reinforced Concrete" for nonsafety concrete
- ACI 349 "Code Requirement for Nuclear Safety Related Concrete Structures" for safety-related structures.

In addition, ACI issues technical publications; among them are the "ACI Structural Journal," the "ACI Materials Journal," and "Concrete International."

What are the ANS standards?

The ANS was formed in 1957 and publishes standards addressing different topics related to nuclear design, plant operation, training, accident analysis and prevention, shielding and confinement, and safety of nuclear power plants. ANS standards of interest to mechanical equipment include:

- ANS-2.27: Criteria for Investigations of Nuclear Facility Sites for Seismic Hazard Assessments
- ANS-2.29: Probabilistic Seismic Hazard Analysis
- ANS-2.23: Nuclear Plant Response to an Earthquake
- ANS-2.26: Categorization of Nuclear Facility Structures, Systems, and Components for Seismic Design
- ANS-2.3: Estimating Tornado, Hurricane, and Extreme Straight Line Wind Characteristics at Nuclear Facility Sites
- ANS-2.10: Criteria for the Handling and Initial Evaluation of Records from Nuclear Power Plant Seismic Instrumentation
- ANS-2.15: Criteria for Modeling and Calculating Atmospheric Dispersion of Routine Radiological Releases from Nuclear Facilities
- ANS-58.2: Design Basis for Protection of Light Water Nuclear Power Plants Against Effects of Postulated Pipe Rupture

How are ASTM standards used in nuclear power plants?

ASTM standards are primarily used in the nuclear power industry for material specifications and material testing. Note that safety-related materials for ASME III plants must comply with ASME II, and therefore the ASTM material must have been included in ASME II if it is to be used. A material listed in ASME II receives the prefix S. For example, the common carbon steel used in many systems in nuclear power plants is the Section II-listed ASTM A106 which is therefore called out as SA106. For nonsafety-related materials, the ASTM designation is sufficient, in this case ASTM A106.

What AWS standards are commonly used in nuclear power plants?

The AWS Foundation was established in 1919 for advancing the application of welding, brazing, soldering, and cutting processes. The common AWS code used in nuclear power is AWS D1.1 "Structural Welding Code," with welding procedures and qualifications as called out in ASME IX and ASME III.

What IEEE standards apply to the qualification of nuclear power plant equipment?

The IEEE was founded in 1884, and provides codes for electrical and electronic systems. Several IEEE standards are used for the design and qualification of mechanical systems and components. The IEEE standards of interest to equipment qualification include:

- IEEE Standard 279 "Criteria for Protection Systems for Nuclear Power Generating Systems"
- IEEE Standard 308 "Standard Criteria for Class 1E Equipment for Nuclear Power Generating Stations"
- IEEE 317 "Electrical Penetration Assemblies in Containment Structures for Nuclear Power Generating Stations"
- IEEE Standard 323 "Standard for Qualifying Class 1E Equipment for Nuclear Power Generating Stations"
- IEEE 334 "Standard for Type Tests of Continuous Duty Class 1E Motors for Nuclear Power Generating Stations"
- IEEE Standard 336 "Installation, Inspection, and Testing Requirement for Instrumentation and Electrical Equipment during the Construction of Nuclear Power Generating Stations"
- IEEE 344 "Standard for Seismic Qualification of Equipment for Nuclear Power Generating Stations"
- IEEE 382 "Class 1 Electric Valve Operators for Nuclear Power Generating Stations"
- IEEE 422 "Guide for the Design and Installation of Cable Systems in Power Generating Stations"
- IEEE 450 "Recommended Practice for Maintenance, Testing and Replacement of Vented Lead-Acid Batteries for Stationary Applications"
- IEEE 494 "Standard Method for Identification of Documents Related to Class 1E Equipment and Systems for Nuclear Power Generating Station"
- IEEE 634 "Cable Penetration Fire Stop Qualification"

What is the meaning of the ANSI designation that we often see in front of a code number, such as ANSI/ASME B31.1?

The ANSI was founded in 1916 by the American Institute of Electrical Engineers, the ASME, the American Society of Civil Engineers, the American Institute of Mining and Metallurgical Engineers, the American Society for Testing Materials, as well as the US. Departments of War, Navy, and

Commerce. ANSI establishes the US rules for the standardization system for public health, safety, business, and technology, which are followed by code-writing bodies such as ASME, AISC, and IEEE. When a code or standard is developed following the rules set by ANSI it is so designated.

Are operating systems treated differently than new designs?

SSCs that are found degraded or nonconforming during plant operation must be evaluated for continued operation, following the NRC Technical Guidance Part 9900 for Operability Determination (OD) or Functionality Assessment (FA) for continued safe operation.

- OD addresses the nuclear safety function of the SSC as described in the Plant Technical Specification (Tech Specs) or FSAR.
- FA addresses the function of the SSCs that may not be safety related.

1.2 CLASSIFICATION

How are SSCs classified in a nuclear power plant?

The various forms of classification of SSCs derive from the fact that a nuclear power plant is designed under the concept of defense in depth. This concept relies on multiple barriers to protect the public and the environment against the release of radioactivity. Typically, there are three physical barriers meant to protect the public, the worker, and the environment: (1) The fuel rod cladding which contains the radioactive fuel, (2) the reactor coolant pressure boundary barrier which cools the reactor core, and (3) the containment barrier.

Is defense in depth different than single failure?

Yes, single failure design means that mechanical and electrical systems have to be redundant, such that the loss of capability of a component to perform its safety function is mitigated by the availability of a second, redundant, component. Defense in depth means that there will be multiple barriers of SSCs to protect the worker, the public, and the environment.

How are codes and standards applied to achieve defense in depth?

Each SSC in a nuclear power plant must be qualified to perform certain design functions: to start-up the plant, to operate the plant, to shut down the plant, to respond to abnormal conditions, and to maintain the plant in a safe condition while it is shutdown.

The classification of events is addressed in several places. First is the SRP, and the criteria of ANS 18.2 "Nuclear Safety Criteria for the Design of Stationary Pressurized Water Reactor Plants." In some plants, events are grouped as Conditions I through IV, where:

- Condition I are events related to normal operation and operational transients
- Condition II are faults of moderate frequency, which may occur during a calendar year
- Condition III are infrequent faults, which may occur infrequently during the life of a particular plant
- Condition IV are limiting faults, postulated but not expected to occur during plant life

In ASME III and in regulations, plant events and their related loading conditions and qualification requirements are classified into four categories referred to as Service Levels. They are:

- Service Level A (previously called normal) events are those that the SSC is subjected to in the performance of its normal service function.
- Service Level B (previously called upset) events are off-normal events, those that the SSC must withstand without damage or repair. Level B events include the operating basis earthquake, and certain thermal transients.
- Service Level C (previously called emergency) events are rare events that can cause large but local deformations that may necessitate removal of the SSC for inspection or repair. ASME III NB-3100 states that for a Service Level C event there should be no more than 25 cycles at a stress larger than the stress that corresponds to 1 million cycles on the fatigue curve of ASME III Appendix I.
- Service Level D (previously called faulted) events are one-time events that can cause general deformations with some consequent loss of dimensional stability and damage requiring repair and even removal. Level D includes the safe shutdown earthquake, and a large loss of coolant accident.

What is the relationship between the design codes and the SSCs classifications in nuclear power plants?

SSCs in a nuclear power plant are classified in accordance with their design function. Their procurement, fabrication, construction, testing, operation, and inspection must meet quality standards commensurate with the importance of the safety function to be performed. SSCs performing a nuclear safety function are classified as Nuclear Safety. SSCs are grouped as

Safety Class 1 (the highest), Safety Class 2, Safety Class 3, and finally Non-Nuclear Safety (NNS). The design codes provide graded requirements as a function of the safety classification of the SSC. The highest requirements apply to Safety Class 1.

How many types of classifications are there in a nuclear power plant?

In nuclear power plants there are three classifications of SSCs: (1) Safety or nonsafety class, on the basis of their function; (2) seismic category or nonseismic, on the basis of their postearthquake function; and (3) essential or nonessential, on the basis of their function following a given postulated pipe break.

What constitutes safety class 1?

10 CFR 50.55a defines what constitutes Safety Class 1; it is the reactor–coolant system (RCS) pressure boundary. Components which are part of the reactor coolant pressure boundary must meet the requirements for Class 1 components in Section III of the ASME B&PV Code, with exceptions spelled out by the NRC in 10 CFR 50.55a. In other words, the reactor vessel, the steam generators (Pressurized Water Reactor - PWR), the reactor coolant pumps, the pressurizer, and the reactor coolant loop piping are safety Class 1. Also, lines attached to this equipment up to isolation valves are Class 1. This includes, for example, the reactor vessel head vent piping located on the reactor vessel head, and the pressurizer surge line.

What constitutes safety class 2 and 3?

Unlike Class 1, 10 CFR 50.55a does not define what constitutes Class 2 and 3 systems. It does say, however, that once an SSC is classified as Class 2 or Class 3, it must meet Quality Group B for Class 2 or Quality Group C for Class 3. The definition of Safety Classes 2 and 3 can be found in RG 1.26 "Quality Group Classifications and Standards for Water-, Steam-, and Radiological-Waste-Containing Components of Nuclear Power Plants," and in NUREG-0800 SRP Section 3.2.2.

What are typical class 2 systems?

Examples of Class 2 systems, in accordance with RG 1.26, are: the RCS pressure boundary excluded from Class 1, the emergency core cooling system; the postaccident containment heat removal, the postaccident fission products removal, the reactor shutdown, the residual heat removal system, the steam generator main steam and feed water up to the containment isolation valve, and the BWR (Boiling Water Reactor) steam from the containment isolation to the turbine stop valve.

What happens if the boundary between class 1 and class 2 fails?

An example that dates back to the 1970s is the rupture of tubes in a PWR steam generator which caused a primary-to-secondary leak of approximately 125 gpm during start-up, when the primary side was at 2233 psi and 491 °F. The primary-to-secondary side leak through the failed tubes caused the primary side pressure to drop by approximately 28 psi/min. The plant and operators responded according to procedures, isolating the damage steam generator, maintaining the PWR primary coolant subcooled (not boiling) while reducing the RCS pressure. These actions successfully reduced the leak from the primary to the secondary side (IN 79-27.)

What are typical class 3 systems?

Examples of Class 3 systems, in accordance with RG 1.26, are: the cooling water and seal water systems to equipment important to safety, and the systems containing radioactive material in sufficient quantity to have an off-site dose consequence in case of leak or rupture.

What are nonsafety systems?

Nonsafety systems are systems which are not Safety Class 1, 2, or 3. They include several systems in the turbine building. They can be viewed as power production systems, such as steam supply to the turbine and the condensate from the turbine. These nonsafety systems are similar in some ways to systems found in nonnuclear power generating plants.

How to recognize the boundaries between the various safety classifications?

The boundaries between safety systems occur typically at isolation valves. These transition points are flagged on Process and Instrumentation Diagrams (P&IDs), sometimes known as Flow Diagrams, and in some plants they are also flagged on piping isometrics. Typically a safety class flag is a triangle containing the numbers 1, 2, 3 and the letters NNS (or a black flag instead of the letters NNS) to indicate the boundaries of safety classes.

Instrumentation impulse tubing, valves, and fittings extending from a root valve in a system, up to and including the last fitting before entering the instrument are classified as Safety Class 2, 3, or NNS, depending on the fluid system classification shown on the system P&ID.

Why does the safety class of an SSC matter?

The safety class of an SSC is important because it controls several attributes: (1) the materials procurement and quality control requirements; (2) the design loads and the level of detail of the design; (3) the quality of fabrication and construction, including the level of NDE; and (4) the amount of ISI and IST that are performed on the SSC during the operating life of the system.

The safety class also controls the level of detail, and the level of review, required in performing ODs during plant operation when a nonconforming or degraded condition is found.

What classifications apply, other than safety class?

Two other classifications apply: Seismic Category, and essential SSCs.

What are the seismic categories?

There are three general Seismic Categories of SSCs. The labeling for Seismic Categories has changed over the years, but they are basically as follows:

- Seismic Category I constitutes SSCs that have a safety function following a postulated earthquake. They are defined in RG 1.29, and classified as Seismic Category I, with a roman number 1, also referred to as SC-I or Seismic Cat.I.
- Seismic Category II/I (sometime referred to as SC-II/I or even SC-II) are SSCs who themselves do not have a postseismic safety function but whose failure could affect an SC-I SSC. We say that the SC-II/I SSC is a source of interaction. These seismic interactions can take one of several forms: (1) falling or swing impact (sometimes called spatial interaction), (2) spray wetting, (3) flood, or (4) system interaction, such as the unintended closure of a valve on an erroneous electrical signal or on loss of air.
- Nonseismic constitutes SSCs that are neither SC-I nor SC-II/I.

Are SC-II/I SSCs designed the same way as SC-I SSCs?

In the assessment of seismic interactions, two questions must be asked:

- Is the seismic interaction credible? In other words would an earthquake cause the failure of the source and would the source reach the target?
- Is the seismic interaction significant? In other words, if the source SSC was to reach the target, would it damage it?

If the answer to both questions is yes, the SSC source is SC-II/I and it must be seismically restrained to prevent the interaction from occurring. But the evaluation criteria for an interaction source do not necessarily have to meet the same stress and load limits as the SC-I target. For example, plastic deformation may be permitted in the source SSC if it will not lead to failure or adverse interaction.

The quality of the calculations and documentation for the SC-II/I source has to be as thorough as the qualification of the SC-I target. For example, in one case, Seismic II/I tanks had been analyzed empty, while the worst case of interaction occurs when the tanks are filled with water. Because of such a mistake, the tanks had to be declared inoperable and an 8-hour action statement was entered for the associated limiting condition for operation (IN 12-01.)

Are there adverse interactions other than seismically induced?

One case of nonseismic interaction that comes to mind happened in the 1980s, when the actuation of the wet fire suppression system caused water damage to safety-related electrical equipment. Plants undertook to review the potential for such water damage, and take corrective action, either adding shields to prevent direct spray, or elevating the equipment base to prevent flood wetting (IN 83-41.)

Are the boundaries between seismic category the same as the safety class boundaries?

No, because while Safety Class boundaries occur at a hydraulic boundary (such as an isolation valve), the SC-I boundary must occur at a structural boundary (for example, a rigid anchor on a pipe). The SC-I SSC must be structurally isolated by an anchor (or an equivalent number of rigid restraints) from the nonseismic SSCs.

What are essential SSCs?

The nomenclature Essential and Nonessential, is applied as part the high-energy pipe rupture hazards analysis. SSCs that must perform a safety function following a pipe break are labeled Essential for that particular break. In summary, for example, a pump may be (1) Safety Class 3 (SC-3), (2) SC-I, and (3) essential (or nonessential) for a given break.

CHAPTER 2

Design Basis Loads and Qualification

Chapter 2 addresses the normal operating loads and postulated accident loads which govern the design of nuclear power plant systems and components. They include normal loads (weight, pressure, thermal expansion), and occasional or postulated loads such as fluid transients, seismic loads, flow-induced vibration (FIV), loads on buried pipe, fatigue, and thermal stratification. We also address in this chapter the concept of leak before break (LBB).

2.1 LOAD DEFINITIONS

2.1.1 Suspended Systems

How many types of suspended systems are there in a nuclear power plant?

Suspended systems, also commonly called distribution systems, are continuous systems supported from walls, ceilings, floors, and steel structures. They are the second "S" in the common acronym SSCs (structures, systems, and components). They are in many ways the same types of systems one would find in any power or process plant, they consist of the following categories:

- Piping and Supports. In a nuclear power plant pipes convey liquids (mostly water) and steam or gases. They are made primarily of stainless steel and carbon steel. They are either Safety or Non-Safety Class, depending on their function, as discussed in Chapter 1. By piping systems we mean the pipes themselves and their fittings (elbows, reducers, tees, etc.) and in-line components (valves, instruments, etc.), and their supports.
- Tubing and Supports. Tubing is piping of a custom size, typically called out by its outside diameter and wall thickness. It is often used for instrumentation and sensing lines, and rarely for process systems. Unlike piping, tubing is not fabricated or procured to ASME B16.10 or B16.19 standard pipe schedules. Tubing in nuclear power plants is typically made of small bore (2 in and smaller) stainless steel, and is either

Nuclear Power Plant Safety and Mechanical Integrity
ISBN 978-0-12-417248-7

welded or joined by specialty fittings which can be dismantled and reassembled for maintenance and instrument calibration.

- Electrical Conduit and Supports. Conduits are tubes containing electrical wiring and cables. They protect long electrical runs of cable and they span from a power source to energized equipment, distribution panels, etc. They are interconnected by junction boxes.
- Cable Trays and Supports. In this application, a bundle of cables spread over flat trays is carried over long distances in the plant, some cable trays are several hundred feet long. Cables inside conduits would land on these trays. Conduit and cable trays are often grouped together and called raceways.
- Heat, Ventilation, and Air-Condition (HVAC) ducting. These are duct systems usually supported by structural steel composed of angle members and bolted steel struts.

What loads apply to the design of piping and tubing systems?

The loading on any given SSC can be categorized in one of several ways: (1) normal load or postulated accident load; (2) service Level A, B, C, or D load; (3) primary load or secondary (self-relieving) load; (4) static (constant) load or dynamic (time-dependent) load; or (5) sustained (applied most of the time) load or occasional (occur once in a while) load.

What are typical static loads?

- Deadweight: The deadweight load includes the weight of the pipe or tubing itself, its contents, the insulation, the in-line components (valves, instruments, etc.), and in some cases a tributary weight of a support that is itself supported in part by the pipe. The deadweight load also includes the water-filled weight of gas or steam lines when they have to undergo a hydrostatic test.
- Internal pressures: We use the plural because a piping system operates at a multitude of pressures, including the test pressure, which is typically larger than the design pressure.
- Thermal modes: We use the plural for this too because there are a multitude of thermal cases that go hand-in-hand with the multiple pressure cases. Thermal loads can in turn be subdivided into two categories: (1) thermal expansion–contraction loads which are global loads on the pipe expanding as a beam structure, including the thermal movements at equipment nozzles, and (2) thermal transients which are local time-

dependent temperature fluctuations of the fluid which in turn causes thermal gradients in the metal.

- Building movements: The pipe or tubing attachment points to equipment or structures can expand due to heating or cooling of the buildings. An example is the normal expansion–contraction of the containment with the seasonal change in external ambient temperature. Another example is the expansion of the containment building or compartment walls due to the heating and pressurization that would result from a postulated pipe break inside containment or in a confined room. Finally, there are the attachment point movements of the supports to the building structures due to a seismic event, known as seismic anchor movements (SAMs). Each one of these types of building movements constitutes a separate loading condition. While SAMs are dynamic in nature, they are analyzed statically by applying the largest differential anchor motions to the system.
- Pseudostatic seismic: These are seismic accelerations ("g" loads) applied statically to the distributed weight of the system, in three directions to represent, in a simplified way, the effect of a postulated earthquake. We will see under what conditions this can be done.
- Pseudostatic wind: Wind loading is addressed in Section 3.3.1 of the Standard Review Plan (SRP). Although a wind load is obviously dynamic, it is typically analyzed as a static wind pressure on the pipe. The primary component of this load is the differential pressure across the pipe, which is based on the wind velocity at elevation height, the shape factor of the pipe, and a wind gust factor.

What are typical dynamic loads?

- Seismic inertia: Nuclear power plants are typically designed for two earthquake levels: the operational basis earthquake (OBE) and the safe shutdown earthquake (SSE). One must assume that during the design life of the plant there will be five postulated OBEs, typically classified as Service Level B (upset) events, and one SSE classified as a Service Level D (faulted) event. We have seen, under static loads, that the seismic building motions are applied statically as SAMs. The second component of a seismic load is the inertial load, where the suspended system deforms due to shaking. The seismic inertial analysis can be performed as a time history analysis or a modal superposition analysis. In the seismic time history analysis the seismic accelerations are imposed to the system as a function of time, and the response of the system is also obtained as a function of time. The time history analysis approach is rarely used in design of piping and tubing systems, as it would require multiple artificial time histories to be generated at each support point, and a detailed model for the time-dependent analysis of the system. Instead, the classic seismic analysis is performed using the modal superposition method, with, as input, an envelope of in-structure response spectra (ISRS) of the attachment points to the structures. The modal analysis of piping systems has to apply several standard rules, which are delineated in each plant's FSAR. These include using the correct damping (R.G. 1.61), the correct modal and directional combination (R.G. 1.92), and the correct rigid range (or zero period acceleration) correction (R.G. 1.92). The inertia response and the SAM response may be combined or treated separately, with the inertia load considered a primary load, and the SAM load, which is displacement limited and therefore secondary, added either to the inertia loads or to the secondary loads caused by thermal expansion.

- Hydraulic transient loads: These are addressed in more detail later, but they typically consist of transients caused by liquid water hammer, water-steam cavitation water hammer, steam hammer, or transients caused by the discharge of safety or relief devices.
- Tornado Load: Tornado loading is addressed in Section 3.3.2 of the SRP. The tornado load is composed of three components: (1) the tornado wind pressure load, as described above for wind but at a higher velocity than regular wind. Reg. Guide 1.76 provides tornado wind speeds ranging from 230 mph for the central United States, down to 160 mph for the western United States. These speeds are subdivided into a smaller translational speed component and a larger rotational speed component. The tornado wind load has to be applied in terms of a moving funnel and is therefore translational and tangential to the target. (2) The second load is the differential pressure caused by the passage of a tornado, an explosion–implosion type load, in which the target is subjected to a pressure difference that can be acting in any direction, depending on the funnel path. A differential pressure in the order of ± 3 psi is the commonly applied load, with a dynamic load factor depending on the natural frequency of the structure. (3) The third load comes from a postulated tornado-generated missile. This is a strike from tornado debris that is lifted and directed with a certain kinetic energy into the target structure. There are standard assumed tornado-generated missiles, such as wood planks, steel rods, pipes, poles, and,

the heaviest, an automobile. Systems that are relied upon for safe shutdown of the plant in case of a tornado strike are designed for these loads. For example, if off-site power is assumed lost and the plant is relying on emergency power supplied by emergency diesel generators (EDGs), then the EDGs must be able to sustain the tornado load or be protected from it.

- Internal missile strike. Internal missile loading is addressed in Section 3.5.1 and 3.5.2 of the SRP. Internal missiles consist of an ejected part from a pressurized component (for example, a valve body-to-bonnet bolt), or an ejected part of a rotating equipment (for example, a pump impeller or blade).
- High-energy (HE) line break loads: These postulated loads, which are unique to the design of nuclear power plants, result from the assumption of a guillotine or side-wall break that causes four dynamic effects: (1) the broken pipe to whip, (2) the discharged fluid to jet onto nearby targets, (3) the pressure of the fluid inside the pipe to blowdown causing a water hammer, and (4) the compartment in which the pipe breaks to pressurize as a result of the fluid discharge in a confined space. These four dynamic effects will be addressed later in more detail.
- FIV: While FIV has been the cause of several instances of fatigue cracking, it is a load difficult to predict at the design stage, that is why it is rarely analyzed at the design stage, relying instead on vibration monitoring during the preoperational test phase in accordance with ASME O&M 3, and periodic vibration inspections.
- Explosions inside components: In high-radiation zones hydrogen can be formed by radiolysis of water contained inside pipes or vessels. The hydrogen could then react with the oxygen in the air to cause a deflagration or a detonation. Another source of hydrogen would result from a severe accident leading to fuel melt. In this case the molten metal reaction in the core releases hydrogen. Two radiolysis-induced detonations occurred in 2001, in BWRs, just weeks apart. In November 2001 a pipe elbow was found ruptured in a BWR residual heat removal (RHR) system with fragments of pipe ejected. This failure was initially thought to be a water hammer but proved to be a hydrogen detonation. The analysis of gases in a sister unit revealed pockets of hydrogen with 46% (volume) hydrogen, 23% oxygen, and 31% steam in the system. Shortly afterward, in December 2001, a containment pressure increase occurred concurrent with an acoustic monitoring signal. This overpressure and loud noise was also traced to a hydrogen explosion inside a

4-inch head spray line, which destroyed nearly 10 ft of pipe. In both cases, the buildup of hydrogen and oxygen gases was caused by radiolysis (IN 02-15). The problem with hydrogen is that its flammability range is quite wide at 4–75% (volume), and the gas mixture requires little energy (fractions of millijoules) to trigger a deflagration and its transition to a detonation.

Can all these loads, static and dynamic, act concurrently or are they treated independently from each other?

The response to this question is in each plant's FSAR, in the form of a table of load combinations, each linked to a Service Level. The load combination tables are typically in Sections 3.8 or 3.9 of the FSAR. Also, the Design Specifications will contain the load combinations required for ASME III components (vessels, piping, pumps, valves, and their supports). Two key considerations that apply when the FSAR and Design Specification load combination tables are written are: (1) the need to follow the load combinations of the SRP Sections 3.8.4 (Seismic Category I (SC-I) structures) and 3.9.3 (ASME III components and their supports) and (2) the need to take into account the type of load as classified in ASME III, either as primary (weight, pressure, seismic, etc.), or secondary (thermal and anchor movements). For piping systems primary and secondary loads are treated separately, with their own stress limits. For supports and for structures they are combined.

Regarding the hydrostatic test pressure, is it set by regulation or by code?

The hydrotest pressure is set by code, namely, ASME III and ASME B31.1. For ASME III piping and tubing systems the hydrostatic test pressure is often $1.25 \times P_{design}$ to which are typically added a 6% code margin and a 75 psi test margin. For B31.1 piping systems the hydrotest pressure is often $1.5 \times P_{design}$ to which is added a 75 psi test margin. In either case, ASME III or ASME B31.1, the hoop stress caused by the test pressure should not exceed yield. For pneumatic testing, the test pressures for ASME III are typically $1.2 \times P_{design}$, and for ASME B31.1 it is $1.25 \times P_{design}$. A 6% margin can be added as in the case of hydrostatic testing. In all cases it is important to check the plant-specific commitment for hydrotest pressure of new systems (ASME III or B31.1) and of in-service systems (ASME XI). In the case of ASME XI, the hydrotest pressure following a repair or an alteration may be performed at $1.0 \times P_{operating}$.

Are the loads on tubing systems the same as on piping?

In principle, the load combinations are the same, but in practice, by design, tubing is not subjected to water hammer and is protected from whip or jet effects; so these large loads would not figure out in design of tubing systems and their supports. Tubing, because of its small size and light weight, is often designed either by static methods, or by using prequalified spans and routing details, sometimes referred to as cookbooks.

There are special zones called break exclusion zones in each plant; What are they and are they designed to the same loads?

Break exclusion zones (BEZs) are sections of piping that penetrate the containment, and in which breaks must be prevented so as to isolate the containment. The BEZs extend from the containment wall to and including the isolation valves outside containment. These sections of pipes, should they break or crack, would provide a flow path from inside to outside containment that cannot be isolated. This of course must be prevented because we do not want what is inside containment to spill outside. So these sections must be prevented from breaking or leaking. The BEZs are therefore treated with great care. This starts with their safety classification, which is as a minimum Class 2, and with the quality of their fabrication and examination. This care continues through the design process with special stress limits, and continues when the BEZ piping is placed in-service through augmented in-service inspection (ISI) under the ASME XI ISI program. These special requirements for the BEZs are listed in the SRP, in Branch Technical Position 3-4.

The BEZ sections of pipe must be protected from any load from piping beyond the BEZ isolation valve. This is achieved by designing a large structural anchor to protect the BEZ isolation valve from loads beyond the anchor. These loads are in the form of weight, thermal expansion, seismic, pipe whip, jet impingement, water hammer, and, for exposed piping, wind and tornado loads. Sometimes, the pipe beyond the BEZ outer anchor is nonseismic, so the structural designer does not have seismic loads that can be used to size the anchor. In such cases, the BEZ anchor can be designed to take a full plastic moment of the pipe, assuming the pipe collapses or buckles at the BEZ interface anchor, in which case these anchors are sometimes labeled plastic moment restraints. The containment penetration is often qualified to protect against pipe loads from inside containment.

What are the design loads for cable tray systems?

The design loads for cable trays are the weight of a full tray (and in some cases an additional live load of 250 lbs) and the seismic load, if the cable trays are SC-I or Seismic Category II/I (SC-II/I). By design, essential cable trays and their supports are protected from dynamic loads such as pipe rupture, jet impingement, missiles, and tornadoes. Thermal expansion–contraction of the trays as the containment temperature changes is usually accommodated by placing flexible joints between the runs of trays.

Regarding seismic design, cable trays terminate into junction boxes or electrical cabinets and equipment, but they are typically connected to these through flexible conduits. SAMs of the equipment are therefore isolated from the cable tray. In seismic buildings, the cables may be tied to trays, to mimic seismically tested configurations.

Were these seismic tests conducted on cable trays?

The seismic performance of cable trays has been the subject of many studies, consisting of reviews of their performance during real earthquakes, and the study of their performance in shake table tests. Seismic tests of cable trays were conducted starting in the 1970s. There are several publications on the subject, including EPRI's "Cable Tray and Conduit System Seismic Evaluation Guidelines" and EPRI's "Longitudinal Load Resistance in Seismic Experience Database Raceway Systems." Another important reference is the EPRI Seismic Qualification Utilities Group (SQUG) "Generic Implementation Procedure (GIP)." The cable tray section of the GIP has a series of practical rules, based on earthquake experience and tests which focus on the key attributes to good seismic performance. These attributes to good seismic performance include the deadweight margin (a deadweight margin of three or more would permit the tray to stay suspended even after large earthquakes), the flexibility of the tray support to base plate attachment, the flexibility of threaded rods from which certain types of trays are suspended from the ceiling, and the strength of the tray support anchorage to the concrete.

What are the design loads for conduit and conduit supports?

Individual conduit runs are typically subject to the same distributed loads as cable trays. Conduits are of course simpler and lighter than cable trays, even when they contain several cables, which is permitted under certain conditions by the National Electric Code.

What are the standards for the design of HVAC ductwork and HVAC supports?

HVAC ductwork is composed of the ducts themselves (round or rectangular), in-line components (dampers to control flow, instruments to measure flow characteristics, heaters, etc.), and the duct supports to concrete or embedment plates. The air handling units with their banks of filters and instrumentation are also an integral part of all HVAC systems in nuclear power plants.

In nuclear power plants, safety-related SC-I ducting and supports are designed to ASME AG-1 "Code on Nuclear Air and Gas Treatment," while nonsafety ductwork is designed to SMACNA with supports designed to American Institute of Steel Construction (AISC) and American Iron and Steel Institute (AISI).

What are the loads that must be sustained for HVAC ductwork and HVAC supports?

First, there is as always the deadweight load (plus usually a typical 250 lbs live load), and the maximum positive and negative differential pressure between the pressure inside the duct and the ambient pressure outside the duct. The worst pressure differential must account for a postulated design tornado condition, and a loss of coolant accident (LOCA) inside containment. As was the case with cable trays, the differential thermal contraction–expansion is accommodated through flexible joints. The other load to be applied in design is seismic, both inertia and anchor motion.

How are HVAC ducts seismically qualified?

Regarding seismic design, the first question is to determine the seismic function of the duct system, and classify this function as one of the following three functional performance categories, from most to least demanding. (1) Operability: does the duct have to perform an active function, which means deliver and control air flow after the seismic event? (2) Leak tightness: does the duct have to remain leak tight, and not permit inflow from or outflow into the room? (3) Position retention: does the duct have to only maintain its position (remain suspended) after the earthquake, but can leak (the seams can open) and does not need to convey air flow?

- In the first case (operability), the duct has an active function, and the ducting system, including fans, power supplies, plenums, dampers and their inflatable seals, the duct itself, the supports, etc., must be seismically qualified.
- In the second case (leak tightness), the duct has to maintain its pressure boundary, the ductwork must remain in position, the joints must remain

tight, and the sheet metal must not tear or punch. For this reason, it is important to design and qualify by analysis the integral attachments between the duct sheet metal and the supports to prevent a tearing or punching of the duct sheet metal. On large rectangular ducts, it becomes necessary to check the stresses caused in the sheet metal by the seismic inertia load, with the sheet metal of the duct considered as a plate.

- In the third and final case (position retention), the duct has to maintain its position, the duct supports must not fail, and the joints must not open to prevent a section of ductwork to fall or unzip. We can see that the level of analysis and testing will vary according to the functional requirement.

Is there a good reference to address seismic qualification of ductwork?

As always, the starting point is the plant FSAR and the corresponding SRP. For ductwork this would be covered under FSAR and SRP Section 3.7.2 Seismic Subsystem Analysis. A good reference is EPRI's "Seismic Evaluation Guidelines for HVAC Duct and Damper Systems," which uses earthquake experience to outline the key attributes for seismic adequacy. These attributes include: duct span, duct supports and bracing, duct stability (not fall-off supports), type of joints between duct sections, attachments, in-line components, duct header-branch differential motions, cantilevered duct sections, stiffeners, and anchorage to concrete.

2.1.2 Systems Supports

What are the types of loads applied on steel or concrete support structures?

The steel or concrete structures supporting suspended systems are typically designed for two general categories of loads: (1) the loads caused by their own weight and (2) the loads imparted by the attached suspended systems.

These loads can be grouped as (1) normal operating loads, (2) test loads, (3) anticipated transient loads, (4) severe environmental loads, and (5) pipe break loads.

Normal loads are loads that will be experienced during construction, normal start-up, operation, and shutdown.

- Dead loads from the structure itself, and from attached components
- Live loads such as movable equipment loads, for example, polar crane loads

- Normal operating thermal expansion or contraction loads

Test loads are generated during preoperational or periodic testing of the structure or attached system:

- Containment pressure test, during which the containment expands by pressure
- Water weight during a hydrostatic test

Anticipated dynamic loads in service:

- Anticipated pressure transients, such as steam or water hammers that are expected to occur from normal system operation, for example, the thrust force caused by safety valve discharge, or the closure of the main steam isolation valve (MSIV) as a result of a turbine trip.

Severe environmental loads caused by postulated natural phenomena hazards:

- Wind loads
- Tornado loads, which are composed of wind loads and tornado missile strikes
- Earthquake loads generated by the OBE or the SSE
- Submergence, waves, and flood caused by a seismically induced tsunami (ocean induced), or seyche (lake induced)
- Flood loads generated by the probable maximum precipitation or overflow of nearby rivers or lakes

Pipe break loads are loads in HE lines:

- Pressure load within or across a compartment walls, generated by the discharge of flashing water or steam from the postulated break
- Environmental thermal expansion following a design basis pipe break, caused by the heating of walls and structures as a result of the discharge of hot water or steam
- Pipe whip reaction loads imparted through the pipe support or whip restraint on the structure
- Jet impingement load on the structure
- Impact load on the structure from a missile generated by a pipe break
- Fluid transient, typically a blowdown of pressure along the system, caused by the pipe break

Which standards are followed for the design of support steel structures?

The supports of ASME III vessels, tanks, piping, pumps, and valves are addressed in ASME III NF. The backup steel structures beyond the NF support boundary are typically designed according to the elastic working stress design methods of AISC or AISC-N690, except for pipe break loads where plastic analysis has been used.

Which standards are followed for the design of HVAC support structures?

HVAC duct supports are typically designed to the following standards:

- ASME N-509 "Nuclear Power Plants Air Cleaning Units and Components"
- ASME/ANSI AG-1 "Code on Nuclear Air and Gas Treatment"
- AISC-N690 "Specification for the Design, Fabrication and Erection of Steel Safety Related Structures for Nuclear Facilities"
- AISI "Specification for the Design of Cold Formed Steel Structural Members"
- SMACNA "HVAC Duct Construction Standards, Metal and Flexible."

Which standards are followed for the design of cable tray support structures?

For conduit and cable tray (raceway) supports, the same AISC and AISI standards would apply, plus standards specific for raceways:

- IEEE "Recommended Practice for Seismic Qualification of Class 1E Equipment for Nuclear Power Generating Stations"
- National Electrical Manufacturers Association, Standard Publication No. VE 1 "Metallic Cable Tray Systems."

Which standards are followed for the design of support concrete structures?

Safety-related concrete structures on the later vintage plants are typically designed to ACI 349 Code. Nonsafety-related concrete structures are designed to ACI 318.

In addition to ACI 349 and ACI 318, there are structure-specific requirements, such as those of NUREG/CR-6608 "Summary and Evaluation of Low-Velocity Impact Tests of Solid Steel Billet onto Concrete Pads" for fuel drop loads in the spent fuel pools.

What standards apply to concrete penetrations?

There are containment components made of steel that are not backed by concrete, for example, portions of penetration sleeves, equipment hatch covers, and airlocks. These are normally designed using the provisions of ASME III article NE-3000, the articles for Penetration Assemblies of ACI 359 "Code for Concrete Containments," and the design limits and loading combinations of Regulatory Guide 1.57 "Design Limits and Loading Combinations for Metal Primary Reactor Containment System Components."

For pipe break loads, the sleeves or forgings of piping and instrumentation penetrations should consider the effects of torsion, bending, and shear loads that the pipe is capable of exerting on the penetration, as limited by the full plastic strength of the pipe and an axial load based on the maximum possible thrust reaction, plus internal pressure from the discharging fluid. If this a designated break exclusion area, the requirements and stress limits for such criteria must be met for stress to be below a certain value as described in NUREG-0800 SRP Sections 3.6.1 and 3.6.2.

What standards apply to concrete liners?

The methods and criteria of ASME III Div.1 Subsection NE "Class MC Components" for metal containments, and ASME III Div.2 "Code for Concrete Containments" (developed by a joint ASME–ACI committee) apply to steel liners anchored to concrete, liner insert plates for piping and duct supports, polar crane and rotating platform insert plates, overlay plates and structural shapes attached to the liner, structural steel attached to the inserts, process piping and instrumentation penetrations, personnel air lock, emergency air lock, as well as equipment hatch anchorage, electrical penetration sleeves, and fuel transfer tube penetration sleeve.

2.2 LOAD COMBINATIONS

Any caution before we start on the subject of load combinations?

Yes. As everything else in this book, in the case of load combinations the engineer must first refer to the plant FSAR and the Design Specifications for safety-related SSCs. We describe here technical guidance and cautions which have evolved over the years, so each plant-specific FSAR should be consulted; oftentimes the load combinations can be found in Section 3.9.3 of the FSAR.

Why is it important to understand loads and load combinations?

The design of SSCs in nuclear plants is founded on two pillars: (1) the classification of the SSC based on its safety function (safety Class 1, highest, then 2, 3, and Non-Safety) and seismic category (SC-I or SC-II/I or nonseismic), and (2) the design loads and load combinations to be applied to the SSC. Load combinations govern the cost of initial engineering, procurement, and construction, and the cost of ongoing engineering and maintenance.

Where are the loads and load combinations defined?

Loads and load combinations are defined for each plant in two documents: the plant FSAR, typically Sections 3.7, 3.8, or 3.9, and for safety-related ASME III components and supports in the ASME III Design Specifications. From there, the load combinations flow down to plant-specific engineering procedures.

Does each plant have its own load combination?

Each plant has its load combination defined in its FSAR, but they comply with the SRP, and therefore they are all similar, even if the symbols for individual loads vary, for example, whether deadweight is labeled DW, W, or some other way. Because of safety relief valve discharges in a suppression pool, there are load sets unique to BWRs.

Are load combinations for supports different from those for the supported components?

Support load combinations are different as a result of the differences between ASME NB/NC/ND which applies to the component itself, and ASME III NF which applies to its support. For example, in designing and qualifying a pressure boundary component, say a vessel or a pipe, the stresses caused by thermal loads (secondary stresses) are treated separately from the stresses caused by deadweight (primary stresses). In the design of the support the thermal and deadweight loads are combined.

The load combinations for component supports must meet Section 3.9.3 of the SRP, and for Class 1 they must also meet RG 1.124 "Service Limits and Loading Combinations for Class 1 Linear-Type Supports," and RG 1.130 "Service Limits and Loading Combinations for Class 1 Plate-and-Shell Component Supports."

Have the load combinations changed over the years?

As we look back, we see key turning points in how nuclear plant SSCs are designed. The first turning point came with the publication of NRC NUREG-1061 in the 1980s.

Many of us remember the development and publication of this important document, since the whole US nuclear industry was asked to comment, and the comments were published. NUREG-1061 was divided into five volumes, each developed by a Task Group of NRC staff. These five volumes, a must read for all nuclear mechanical and structural design engineers, are:

- Volume I published in 1984, entitled "Investigation and Evaluation of Stress Corrosion Cracking in Piping of Boiling Water Reactor Plants," addressed the causes and remedies to the occurrence of stress corrosion cracking (SCC) in the BWR primary system.
- Volume II published in 1985, entitled "Evaluation of Seismic Designs - A Review of Seismic Design Requirements for Nuclear Power Plant Piping," addressed conservatisms in design, the two-earthquake design approach (OBE and SSE), damping values, spectra modification, multi-input response spectra, nozzle loads, support design, inelastic analysis, and snubbers, and nuclear engineering practices around the world related to these topics.
- Volume III published in 1984, entitled "Evaluation of Potential for Pipe Break," addressed the potential for pipe break, how the industry postulates breaks in HE lines and cracks in moderate-energy lines, and the effects of these on design. This document opened the door to the elimination of arbitrary intermediate breaks and to the adoption of the LBB concept to eliminate the dynamic effects of a postulated LOCA.
- Volume IV published in 1984, entitled "Evaluation of Other Loads and Load Combinations," addressed six areas: (1) the combination of SSE and LOCA, (2) seismic response for piping supported off different structures, what is sometimes called independent support motions, (3) plastic strains and effects of dynamic strain rates, (4) design for water hammer, (5) the prediction of safety and relief valve loads, and (6) how to consider vibration loads. Of direct importance to design is the first topic, the combination of SSE and LOCA which causes very large, design-controlling loads on the reactor coolant system, the reactor vessel, the vessel internals, and the fuel assemblies. The matter that was debated in the 1970s and captured in Volume IV was this: if an SSE could cause the primary system piping to break in a guillotine manner, with full separation, causing an LOCA, then the LOCA load should be added to the SSE. On the other hand, if the reactor coolant system is seismically designed to prevent an SSE-induced break, why should SSE and LOCA loads be combined? This is an important point because it affects the design of the reactor coolant system, the reactor pressure vessel, and

the fuel assemblies. The combination of LOCA and SSE has been the subject of much discussion in the 1970s. Some of this discussion can be revisited by reading two NUREGs published in 1980, NUREG-0484 "Methodology for Combining Dynamic Responses" and NUREG-0609 "Asymmetric Blowdown Loads on PWR Primary Systems."

- Volume V, published in 1985, entitled "Summary - Piping Review Committee Conclusions and Recommendations" captures the conclusions and recommendations from Volumes I through IV.

Are SSE and LOCA loads combined by absolute sum?

NUREG-1061 concurred with an earlier recommendation dating to 1978 (NUREG-0484) to combine LOCA and SSE by square root sum of the squares (SRSS) rather than by absolute sum for the reactor coolant pressure boundary. In 1980, Revision 1 of NUREG-0484 extended the SRSS of LOCA and SSE loads to SSCs beyond the reactor coolant system. However, some plants were already designed for an absolute sum of SSE and pipe break. There was a third set of plants that subdivided the LOCA loads into a transient short-term load (whip, jet, and initial blowdown) that is combined SRSS with SSE, and a steady-state load (steady-state blowdown discharge through the break) that is added by absolute sum to the SSE.

What was the second key turning point in load combinations, beyond NUREG-1061?

The second turning point is more recent, and in fact, because it is so recent, it is not implemented in the operating fleet in the United States, but only in the plants under construction at the time of this writing, the combined license plants. It is the elimination of OBE, under certain conditions and with certain restrictions. The first step toward this approach came in 1993, with the streamlining of regulatory requirements for new reactor designs (evolutionary and passive designs), which was documented in SECY-93-087 Policy, "Technical, and Licensing Issues Pertaining to Evolutionary and Advanced Light-Water Reactor (ALWR) Designs." The Commission approved the elimination of OBE by memorandum July 21, 1993. The position was finalized in 10 CFR 50 Appendix S "Earthquake Engineering Criteria for Nuclear Power Plants."

The elimination of OBE from the design of SSCs was driven by a practical fact. Plants have always been designed for two earthquakes: a very large earthquake, the SSE that is postulated to occur once in the plant design life, and five smaller Operating Basis Earthquakes (OBE), more likely and for which the plant should be able to continue to operate. The SSE is a Service Level D (faulted) event and the OBEs are Service Level B (upset) events. But, because of the way stress limits are specified in the design codes, the OBE, the smaller earthquake, oftentimes controlled the design; it had a lower demand than SSE but a capacity (allowable stress) that was reduced to a larger proportion than the reduction in demand. This made meeting OBE more difficult than meeting SSE, and caused iterations in design to qualify the OBE, even when the SSE was already qualified.

The industry and the NRC concurred that there should be no need to explicitly design for the OBE if the following conditions were met (10 CFR 50 Appendix S):

- The OBE is defined as sufficiently small, the agreed-upon level being an OBE at one-third of the SSE
- The plant commits to shutdown and inspect its SSCs for any sign of damage in case of an earthquake which exceeds one-third of the SSE
- An earthquake half the SSE would still be used for Class 1 fatigue analysis and IEEE-344 and ASME QME-1 seismic qualification testing

What Is the general form of load combinations?

The load combination for each plant is clearly stated in its FSAR and, for safety class components and supports, in the Design Specification. If the OBE is not eliminated, the load combinations are typically in the following, simplified, form:

Design Condition

- $P_{design} + DW$

Test Condition

- $P_{presssure\ test} + DW_{test\ fluid}$

Service Level A

- $P_{SLA} + DW + FT_{SLA}$; and TH_{SLA}

Service Level B

- $P_{SLB} + DW + FT_{SLB\text{-}OBE} + OBE$; and
  $P_{SLB} + DW_{deadweight} + FT_{SLB}$; and TH_{SLB}

Service Level

- $P_{SLC} + DW + FT_{SLC}$; and TH_{SLC}

Service Level D

- $P_{SLD} + DW + FT_{SLD\text{-}SSE} + SSE + DBPB$; and
  $P_{SLD} + DW_{deadweight} + FT_{SLD}$; and TH_{SLD}

In this simplified format, we use the following nomenclature: P_{design} is the design pressure; DW is the deadweight load; $P_{pressure\ test}$ is the pressure during the hydrotest or pneumatic test; $DW_{test\ fluid}$ is the water-filled piping system if hydrotested; P_{SLi} is the pressure for the Service Level i where i is A, B, C, or D; and FT_{SLi} is the fluid transient for a Service Level i. If the transient is caused by the OBE or SSE then it is noted as $FT_{SLi\text{-}OBE\ or\ SSE}$ otherwise it is noted simply as FT_{SLi}; OBE and SSE are the OBE- and SSE-induced inertia and anchor motion loads; DBPB are the loads caused by the postulated HE pipe break (whip, jet, blowdown, and subcompartment pressure), and if the break is in the reactor coolant system, then DBPB is an LOCA; TH_{SLi} are the various thermal expansion–contraction loads for Service Level i.

A few words of clarification are in order:

- Each plant has its own load combination, and its own nomenclature. In most plants the load combination tables are longer because they spell out the individual transients that we grouped here under the single heading FT.
- In some cases, the DBPB is subdivided into a transient short-term load (whip, jet, initial blowdown) that is combined SRSS with SSE and a steady-state load (steady-state blowdown) that is added by absolute sum to the SSE.
- The thermal expansion–contraction loads are listed separately because for pressure boundary components (vessels, piping, pumps, and valves) they are analyzed separately as secondary stress, but for support structures the thermal loads, including friction during thermal motion, are added to the primary loads.
- For ASME III Class 1 components, the thermal loads are not only those caused by structural expansion and contraction loads but also those related to through-wall stresses caused by thermal transients, that are combined with the other loads in ways described in ASME III NB.
- Outdoor systems must be designed for natural phenomena hazards other than seismic. These include wind, tornado, ice, and snow. In some cases, near volcanoes, the design also includes ash fall.

Is the design pressure the largest pressure a system will sustain?

Not in nuclear power plant design. ASME III permits that the pressure exceed the design pressure for Service Levels B, C, and D. For example, the Service Level D pressure is permitted to be twice the design pressure in a pipe, and 1.5 times the design pressure in a valve, under certain conditions listed in ASME III.

An interesting question was raised in the late 1990s regarding thermal relief of water trapped between closed isolation valves. It can be readily calculated that the pressure inside a trapped section of pipe will quickly rise when the containment temperature increases to approximately 300 °F following a postulated pipe break. This question was raised as GL 96-06 and addressed in EPRI Report TR-108,812, "Response of Isolated Piping to Thermally Induced Overpressurization during a LOCA (GL96-06)."

2.3 DYNAMIC LOADS

2.3.1 Fluid Transients

By fluid transient do we mean water hammer?

The term "fluid transient" is used to describe the full range of water hammers, whether they are caused by liquid water (as in "water hammer"), by steam, or by two-phase flow (liquid and steam, or liquid and gas such as air). In a nuclear power plant, fluid transients occur in water and steam systems, so they can be labeled water hammer; but in other industries fluid transients can occur in fluids other than water or steam, you could have a pressure transient in an acid line, an acid hammer, or a pressure transient in a crude oil pipeline, an oil hammer.

Water hammer was recognized early on as a challenge in nuclear power plant operation, and was classified as an Unresolved Safety Issue (USI) A-1. With the high-pressure, high-temperature systems in nuclear power plants, and the large number of water and steam systems, water hammers cannot be categorically eliminated, but through good engineering, maintenance practices, and operations, they can be reduced to the point where they do not pose a safety risk. Part of this effort is to advertise the water hammers that do occur, their causes, and their prevention as has been done by the NRC, utilities, EPRI, NEI, etc. (IN 85-76).

How many types of fluid transients are there?

There are several ways to classify fluid transients. One way is to group them by their fundamental physics. In this case, we would have three major classes of fluid transients:

- The first class of fluid transient is the single-phase transient, which involves a single phase (liquid, gas, or steam) and is caused by a rapid change of flow velocity which results in a rapid change of pressure. This is the most straightforward type of transient, since it can be readily quantified. For example, if a valve in a water system closes more quickly than a critical time, the upstream side will undergo a water hammer, with a pressure rise that can be calculated by a simple formula, and the downstream side will undergo a similar reduction in pressure.

- The second class of fluid transient is the liquid–vapor transient, in which the vapor phase can condense into the liquid phase, and vice versa. These transients are caused by a vapor bubble mixed in the liquid system. As the vapor bubble cools down and eventually collapses, it causes a bubble collapse water hammer. This transient is also referred to as cavitation water hammer. On the other hand, when hot pressurized fluid enters a low-pressure region, for example, due to the opening of a valve, the fluid will flash into vapor bubbles. This bubbling of steam in water has caused significant vibrations.
- The third class of fluid transient is the liquid–gas transient in which the two phases, the liquid and the gas, do not convert into each other; the liquid stays liquid, and the gas stays gas during the transient. What happens here depends on the relative quantity and velocity of the liquid and gas phases. Different flow regimes can exist, some harmless and others not so. For example, in a stratified flow, the gas will flow over the liquid phase. In a plug flow, the gas will be swept at the much lower velocity of liquid, which is typically in the range of 0–10 ft/s. In a slug flow, a slug of liquid is propelled at the velocity of gas, or steam, which is much larger than the typical liquid flow velocity, in the order of 100 ft/s or more. The propelled slug or slugs of liquid will then impact bends, partially open valves, tees, etc., causing dynamic loads in the system.

Do water hammers occur from day one, during preoperational testing?

Not necessarily. Most conditions leading to water hammer will only happen under certain operating conditions, and therefore systems can operate for a long time without water hammer, until one particular valve alignment, or flow regime, or water level is attained. One case in point is a water hammer that occurred in an auxiliary feed water (AFW) system. The first sign of water hammer came as the result of an inspection which revealed that eight hangers were loose or damaged on the AFW piping. This was unexpected because operating limits had been imposed purposely to prevent water hammer. As a result, the system had operated without water hammer for many years. But then the configuration of the system was modified, and lines were rerouted. In the original configuration, where water hammers had been prevented, the AFW system discharged into the main feed water (MFW), and from there to the steam generators (SGs) through the MFW spargers. In the modified configuration, the AFW line fed directly into the SG, with its own spargers. This new system layout was duly tested, with no evidence of water hammer. But as the water level in the SG fluctuated, the AFW spargers operated alternatively covered and uncovered, which led to bubble collapse water hammer (IN 84-32).

How are these fluid transients accounted for at the design stage?

In the design of nuclear power plant piping systems, when it comes to predicting and designing for transients (water hammers) it is useful to consider three possible cases:

- Anticipated transients: These are transients we know will happen, by design. A simple case would be the thrust force caused by the opening of a safety or relief valve. Another example, common in many pressurized water reactors (PWRs), is the discharge of the pressurizer safety valve in which the discharging steam will entrain a slug of water down the discharge pipe. Another classic example is the closure of an MSIV that will suddenly cutoff the flow of steam to the turbine. This rapid flow stoppage of large quantities of steam, in lines 24 inch and larger, causes a steam hammer in the system. Pump start-up or trip or check valve slam or filling of a voided line are among some examples of transients that are unavoidable and are known at the design stage and therefore must be analyzed and mitigated as part of the design process. The layout of the lines and their support arrangements will have to account for these effects. For example, piping systems would have to be properly vented by design to minimize vapor or gas void accumulation inside the pipe prior to pump start.
- Unanticipated transients: These are transients that may occur during the plant lifetime and which are the result of an error in operation or maintenance, or are caused by a component failure. An example would be the accumulation of condensate in a steam line due to a faulty steam trap or as a result of an incomplete drainage of the line after a shutdown. The introduction of steam in the presence of this accumulated condensate could cause a condensation-induced, bubble collapse, water hammer. By nature these unanticipated events cannot be predicted at the design stage, and therefore cannot be reliably analyzed. They have to be prevented. In nuclear power plants this is achieved through rigorous operating and maintenance procedures and practices, and frequent ISI and in-service testing of components.

- Postulated transients: These are in some ways unique to the nuclear industry. They are events that are not anticipated to occur but must be analyzed as mandated by safety regulations. A classic example is the blowdown caused by a postulated pipe break in a high energy (HE) line. The postulated break causes the fluid pressure to drop instantaneously at the break, and a depressurization wave to travel down the pipe on both sides of the break as the pipe blows down its contents. Such a pressure wave will cause unbalanced forces in the system.

Regarding the anticipated transient caused by an MSIV closure, How is this transient relieved?

Regarding MSIV closure on full flow, this transient typically causes the largest pressure spike in the main steam system, and is relieved by a bank of safety valves placed just upstream of the MSIVs. This whole transient is analyzed in much detail, and tested, because it causes large pressure spikes in the main steam system, and high flow rates through the safety valves. Ear plugs are essential when testing the MSIVs to closure. The discharge flow of steam through the safety valves causes pressure drops in the piping upstream of the valves, in the order of 100 psi over 20 ft of upstream 6 inch pipe. This upstream pressure loss must be modeled in sizing the safety valves to check that they will not chatter as the pressure drops, and to limit the maximum accumulated pressure in the main steam header (IN 97-09).

Main steam safety valves are sometime tested in situ, with the insulation removed at the ambient air temperature. In-shop tests with steam are sometimes performed with insulation, at a controlled temperature of 140 °F. Interestingly, these two tests, at different ambient temperatures, result in different valve lift pressures, with the insulated hotter valve lifting more easily, within the pressure set point tolerance of −3% to +1%, while the uninsulated cooler valve lifts above the set point tolerance (IN 96-03).

In the category of anticipated transients, since they must be designed for, can more guidance be provided regarding which systems or events are typically analyzed for water hammer?

It depends on the type of reactor, BWR or PWR, and the operation of particular components: how fast valves close, how quickly pumps start, etc. It is therefore a decision that must be made case by case. For the purpose of illustration, in a PWR, the following systems may see anticipated water hammers that would be analyzed at the design stage:

- Relief Valve Discharge: Chemical and Volume Control System; Residual and Heat Removal System; Main Steam; Pressurizer; Component Cooling Water
- Pump Start-up or Trip and subsequent check valve closures: Containment Spray; Safety Injection; Service Water
- Turbine Trip: Main Steam; Feedwater; AFW

How are these transients analyzed?

Going back to the three classes of transients, the type of analysis will depend on whether the transient is (1) a single-phase liquid or steam system, (2) a two-phase liquid–vapor bubble collapse, or (3) a liquid–gas slug flow. There are basically two families of software packages for the analysis of fluid transients:

- The system software that simulates the network of pipe segments and solves the fundamental single-phase or two-phase hydraulic equations along the lines.
- The computational fluid dynamics (CFD) software that can solve the transient and, if necessary, the fluid–structure interactions in a more precise manner, but for a limited volume.

In the case of a single-phase transient there are several system hydraulic simulation software that can model and calculate the transient pressure in the line and the subsequent packing of the line as the pressure wave travels and reflects at changes of flow area or at closed ends. In the case of two-phase flow, things are more complicated. There are fewer system software that can correctly simulate the two-phase transient, and in some cases CFD analysis may be necessary.

Can you provide words of caution for those embarking on a fluid transient analysis?

- First, you must understand the class of transients being analyzed (single phase, liquid–vapor, or liquid–gas).
- Second, obtain the correct initial conditions of the fluid, marked up on a P&ID showing valve alignment, flow rates, pressures, and temperatures at the onset of the transient.
- Third, try to estimate by closed-form solutions the order of magnitude of the expected pressures and forces to be able to later verify the general validity of the numerical simulation results.

- Fourth, keep in mind that what starts as a single-phase transient can quickly turn into a two-phase transient. A classic example is when the pressure rise in a water system caused by rapid valve closure turns into an equal and opposite pressure drop as the pressure wave reflects at a closed end; and if the pressure drop is larger than the initial pressure some of the liquid will vaporize. Another case in point is when the pressure drops sufficiently to separate air entrapped in the liquid, forming a liquid–gas system.
- Fifth, use software that has been benchmarked against experiments, or that can be verified and validated against closed-form solutions.

With this information in hand, How should the hydraulic transient analysis process proceed?

- First, identify which lines will be subjected to anticipated transients and to postulated transients, as we defined above, and write in each case a clear definition of the transient sequence.
- Second, gather the piping system data and initial hydraulic parameters.
- Third, select the software to be used based on the type of transient and verify its validity using simple models that can be benchmarked.
- Fourth, model and run the hydraulic analysis and obtain output time histories of pressure and other parameters such as flow rates. Some software will also automatically calculate the hydraulic force time histories by multiplying pressures by flow areas.

At this point, the hydraulic analysis output, in the form of pressure vs time or force vs time, is turned over to the stress analyst who will apply these pressure time histories or force time histories to the model to obtain loads, movements, and stresses throughout the system. These are combined with the concurrent loads, and compared to allowable limits, in accordance with the plant safety analysis report.

Can you provide words of caution for performing the structural transient stress analysis?

The first piece of advice that comes to mind is to apply the hydraulic forces dynamically, as time histories. Except in the simplest of cases, a static analysis may not yield the correct results. Applying the loads statically can be unconservative because it can cause loads on separate segments to partially or fully cancel each other. Also, by ignoring the timing of the loads the analysis misses the dynamics of the phenomenon and the relationship between the driving force frequency and the frequencies of the beam modes of the system. Also, a static analysis ignores the impulsive nature of the hydraulic loads. The simplest of cases that can be statically analyzed are loads such as a relief valve discharge loads on a short downstream discharge pipe; an example of such analysis is provided in an Appendix in ASME B31.1.

A second important point is to verify the accuracy of the structural piping system model and its ability to capture and respond to the high-frequency content of the hydraulic loads. This involves having sufficient nodes in the model and having a sufficiently high cutoff frequency in the solution routine.

A third important point comes to us from observing the damage in piping systems in actual water hammers: (1) the beam effect damage (the pipe jumps, supports are deformed, anchor bolts are pulled out of concrete, etc.), and (2) the hoop effects (overpressure causing components to break, flange gaskets to blow out and leak, instruments to exceed their range and be damaged, etc.). The pipe stress analysis only addresses the response of the piping system as a beam. A separate evaluation needs to be conducted to check the pressure rating of components and fittings (valves, flanges, instruments, etc.), and gaskets and packing against the peak pressures generated during a transient.

So far we have focused on fluid transients in the design process; How about fluid transients, water hammers, that happen in operation?

What was said above applies to transients that occur in service: The importance of understanding the nature of the transient (single phase, two-phase liquid–vapor, two-phase liquid–gas), the cause of the transient, the initial conditions at the onset of the transient, the correct hydraulic and structural analysis of the event, etc.

If a transient does occur in the plant, the plant will follow its condition report and investigation process. What follows are some technical suggestions in root cause analysis, prevention, and mitigation.

- Understand the initial operating mode, and valve and pump lineups, which is done best when color coded on a P&ID.
- Gather system parameters at the onset of the transient, and identify what particular mode of operation was taking place (stop or start of a pump, opening of a valve, discharge of a relief valve, flow rates in and out of tanks and vessels, plant start-up or plant trip).
- Gather the system parameters (pressures, flow rates, temperatures, liquid levels, etc.) from the control room.

- Determine which class of transient took place (single-phase liquid or steam, two-phase liquid–vapor or liquid–gas), and develop the possible transient scenarios. Obtain operations and system engineering agreement that the transient scenarios outlined are likely.
- Numerically simulate the transient to verify that the structural damage that has occurred matches with our understanding of the thermohydraulics.
- Simulate the transient on the plant simulator to verify that the system response is as estimated.
- Closely inspect the system for damage to the pipe, insulation, supports, attached equipment, valve internals, pump impellers, etc.
- Repair or replace damaged piping, components, equipment, and supports. Upgrade their design if necessary.
- If the transient was caused by an error in operation, a shortcoming in maintenance, an inadequate component, or a component malfunction, take preventive measures (hardware changes or improvements to maintenance practice or operating procedures) to prevent recurrence. If the transient was caused by an anticipated event, and if there is no means to prevent its recurrence, in other words, it should be considered a design basis event, then analyze the transient and take measures to reduce its magnitude, so as not to repeat the damage.

What sources of information are available to learn more about fluid transients?

There are so many textbooks on the subject, even a partial list is not possible. We will only mention two references: The first reference was compiled by EPRI and written with nuclear power plant engineers in mind; it is the *Water Hammer Handbook for Nuclear Plant Engineers and Operators*. The second is the classic 1955 *Waterhammer Analysis* by John Parmakian, a well-written book that highlights the fundamental concepts.

Industry Operating Experience databases should be consulted for water hammer, as well as EPRI report NP-6766 "Water Hammer Prevention, Mitigation, and Accommodation" which captures plant experience with water hammer. The NRC has also published several important documents on the subject: Information Notices 85-76, 87-10, and 91-50, as well as Generic Letter 96-06 "Waterhammer Issues Resolution. Technical Basis Report," NUREG-0291 "An Evaluation of PWR Steam generator Waterhammer," NUREG-0918 "Prevention and Mitigation of Steam

Generator Water Hammer Events in PWR plants," NUREG-0927 "Evaluation of Water Hammer Occurrence in Nuclear Power Plants Technical Findings Relevant to Unresolved Safety Issue A-1," NUREG-1061 Volume 4 "Evaluation of Other Loads and Load Combinations," and NUREG/CR-5583 "Prediction of Check Valve Performance and Degradation in Nuclear Power Plant Systems Wear and Impact Tests."

2.3.2 Seismic Design

2.3.2.1 Deterministic Analysis

Is seismic design required for all SSCs in a nuclear power plant?

Seismic design is required for all SC-I SSCs and for II/I sources of interaction. This is typically most SSCs inside the reactor building and SSCs in many parts of the auxiliary, and a few systems in the turbine buildings. More than any other industry, seismic design permeates much of the design of a nuclear power plant. The massive walls and steel structures, beams, base plates, and support frames are a direct result of seismic design, as shown in Figure 2.1. So are the large number of dynamic restraints in a nuclear power plant, Figure 2.2.

Figure 2.1 Seismically designed tube steel.

Figure 2.2 Seismic dynamic restraints are ubiquitous in a nuclear power plant.

Using a broad brush, what is the seismic design process for SSCs?

The seismic design starts with the regulation that mandates that nuclear power plants be capable of sustaining earthquakes without damage to the core or containment. This is embodied in 10 CFR 50 Appendix A General Design Criterion (GDC) 2 "Design Bases for Protection Against Natural Phenomena," and Appendix S "Earthquake Engineering Criteria for Nuclear Power Plants." These two upper-level requirements flow down throughout Chapter 3 of the SRP, and its corresponding Regulatory Guides (RG).

The seismic design process can be summarized as follows:

- Seismologists and geoscientists reach a consensus on the probabilities of potential sources of earthquakes in the region around the plant, and they define the seismic source terms, the possible epicenters, and associated seismic magnitudes.
- The geoscientists then develop the attenuation functions that describe how the seismic waves will travel from the epicenter to the site.
- The site soil layers are then characterized from the ground surface down to the bedrock, to predict how the seismic waves will travel from the bedrock to the ground surface.

- The buildings and soil are modeled in a combined model, the soil–structure interaction (SSI) model, to then develop the seismic motion at the building foundations.
- The seismic motions at the building foundations are applied to a finite element model of the building. This will be used to, on one hand, analyze and qualify the building structure, and on the other hand develop ISRS at the various floors and walls in the building. The ISRS will then be used to design SSCs inside the building.
- The ISRS are then used for qualification by shake table tests, or the ISRS are applied to models of the SSCs to calculate their seismic response in terms of movements, loads (forces and moments), and stresses. These results are then combined with the other concurrent loads to design the SSCs and their supports.

Where is the likelihood of earthquakes defined?

In a deterministic seismic analysis, which is required for all plants, the plants are designed for specified earthquakes: The Operating-basis earthquake (OBE) and the Safe shutdown earthquake (SSE) ground motion. These are defined in 10 CFR 50 Appendix S:

- *The OBE is* the vibratory ground motion for which SSCs necessary for continued operation without undue risk to the health and safety of the public will remain functional. In practice, plants are designed to sustain five postulated OBE's, unless they are exempted from explicit OBE design as discussed earlier. The OBE is typically classified as a Service Level B event in the load combination table of SRP Section 3.9.
- *The SSE is* the vibratory ground motion for which certain SSCs must be designed to remain functional. In practice, plants are designed to sustain one SSE. The SSE is a Service Level D event, with certain SSCs being capable of performing their safety function after the SSE. RG 1.165 "Identification and Characterization of Seismic Sources and Determination of Safe Shutdown Earthquake Ground Motion" defines the SSE as having a median probability of occurrence of 1 in 100,000 per year.

How are SSCs seismically qualified?

A good discussion on qualification of SSCs is given in ASME QME-1 "Qualification of Active Mechanical Equipment Used in Nuclear

Facilities." We will address two aspects that are essential to a correct seismic qualification process.

- Seismic Function: The need to assign a seismic function to each SSC that must be qualified. The method of qualification will depend on the seismic function. There are three possible seismic functions:
  - Operability (or active function) applies to components that must change state during or after the earthquake. This is the case for all SC-I electrical equipment and instrumentation and controls. It is also the case for certain SC-I mechanical equipment such as pumps, compressors, fans, and valve operators.
  - Pressure boundary integrity (or leak tightness) applies to pressure equipment and systems (tanks, vessels, pumps, valves, pipes and tubes, ducts, etc.) that must remain leak tight to prevent loss of contents, prevent flooding, and prevent spraying.
  - Position retention (or structural stability) applies to all SSCs that cannot fall, overturn, tip, or swing to the point of impact.
- Qualification Method: The correct method of qualification depends on the type of SSC and its seismic function. There are basically four methods of seismic qualification:
  - Qualification by testing using a shake table, which is common to qualify the operability of (1) electrical equipment, their components, instruments, and controls, and (2) active mechanical equipment. In a few cases the test can be static, such as applying a static lateral force to a valve operator to check the strength of the valve-operator-to-body connections.
  - Qualification by analysis, which is common to qualify distribution systems (piping, tubing, HVAC ducts, conduit, cable trays), pressure equipment (tanks, vessels), equipment supports, and anchorage. The analysis can be dynamic (most often modal analysis) but also, under certain conditions, it can be performed by statically applying the seismic loads in each direction.
  - Qualification by natural earthquake experience data. This approach consists in comparing the SSC to be qualified to the performance of similar SSCs in natural earthquakes. The method of earthquake experience data has been developed in the 1980s by the Seismic Qualification Utilty Group (SQUG), and implemented starting in the late 1980s and in the 1990s to address what was at the time NRC unreviewed safety issue USI A-46. The method is recognized by ASME QME-1 and IEEE-344, and is documented in the SQUG

Generic Implementation Procedure (GIP). It can only be applied for certain plants, and under certain licensing conditions.

- The final method of seismic qualification is self-evident, it is the qualification of an SSC based on the previous qualification of a dynamically similar SSC. The challenge with this fourth approach is to establish a close dynamic similarity between the reference SSC and the SSC to be qualified, this question of dynamic similarity is addressed in 10 CR 50 Appendix S, ASME QME-1, and IEEE-344.

Must seismic qualification by analysis be always a dynamic analysis?

The seismic analysis of systems and components in nuclear power plants is typically performed by dynamic analysis, in particular modal response spectra analysis. This is unlike all other industries where seismic analysis is based on static methods defined in standards such as American Society of Civil Engineers (ASCE)-7. The static method can also be used, under certain conditions spelled out in SRP Section 3.7.2 "Seismic System Analysis":

- The system must be simple in shape and realistically represented by a simple model and the static method must produce conservative results.
- The static analysis method must account for the relative motion between all points of support.
- The static seismic load in each direction must be based on 1.5 times the peak acceleration of the ISRS, but a factor less than 1.5 may be used with justification.

What advise can you provide to those performing the seismic analysis of distribution systems?

The work should be done to a correct and well-written procedure which addresses the following:

- The scope and seismic functional requirement of the SSC to be qualified. The scope is best defined by drawings, and the functional requirements should be one of the three defined above (operability, leak tightness, position retention).
- The applicable loads and load combinations, as discussed earlier in this chapter.
- The codes and standards to be applied, which will specify the equations and stress, load, or other limits

- The SRP and RG requirements to be followed: this includes several sections of SRP Chapter 3, as well as RG 1.61 for damping, and RG 1.92 for modal and directional combinations.
- The quality of the analytical model and its degree of refinement, the boundary conditions, the stiffness, eccentricities and tributary weight of the supports, and the coupling or decoupling of the attached equipment or structures.
- The qualification requirements, which typically include stress limits, load limits on attached equipment nozzles, loads on supports, acceleration on valve operators, movements against clearances, local stresses at welded attachments, and movement or load limits at mechanical joints.

The analysis must meet certain tenets of engineering excellence: It must be correct, complete, clear, concise, consistent in format, compliant to the licensing basis of the plant, and cost effective, the 7 c's.

Are SC-II/I SSCs treated differently than SC-I SSCs?

Yes. The approach to seismic interaction (SC-II/I) differs in some ways from the approach to seismic qualification of SC-I SSCs. In a II/I evaluation you must first ask yourself whether the interaction is (1) credible (can the source of interaction reach the target?) and (2) if the interaction is credible, is it significant (will the source damage the target's seismic function)? The qualification criteria for a II/I source of interaction can be more limited than the full qualification of an SC-I item. For example, if the concern is limited to a II/I spatial interaction by falling but there is no spray concern, the source may be allowed to exceed the stress limits of its pressure boundary provided its supports and anchorage do not fail. Oftentimes, a plant will have an analysis and qualification procedure specific to SC-II/I seismic interactions.

2.3.2.2 Probabilistic Analysis

What is a seismic probabilistic risk assessment?

A seismic probabilistic risk assessment (SPRA) is basically the study of what would happen if an earthquake larger than the design basis SSE would occur. It is probabilistic in the sense that instead of specifying the magnitude of this beyond-design basis event (BDBE), a whole suite of earthquake magnitudes, from small to large, are studied, each magnitude with its probability of occurrence or exceedance: Obviously, low-magnitude earthquakes have a larger probability of occurrence than the larger and more devastating earthquakes.

What was the impetus behind the SPRA work that started in 2011 in the United States?

Following the Fukushima accident which occurred as a result of the Great Tohoku Earthquake of March 11, 2011, The USNRC issued several reports and actions, including the following key documents:

- Fukushima Near-Term Task Force (NTTF) report in July 2011, with several recommendations for US plants
- NRC SECY-11-0124, September 2011, recommended actions without delay
- NRC SECY-11-0137, October 2011, prioritized the post-Fukushima action
- NRC SECY-12-0025, March 2012, authorizes 50.54(f)
- Letter 50.54(f), March 2012, to all US operating plants, including a timeline for actions. Recommendations 2.1 and 2.3 applied to the seismic probabilistic risk analysis (SPRA).

In summary, these documents requested an in-depth investigation of each plant's capability to safely shutdown and remain safely shutdown and cooled in case of a BDBE earthquake, in other words, an earthquake larger than the SSE.

As a result, the US plants, coordinating their efforts through EPRI, developed the following two implementation documents, which were accepted by the USNRC, and the plants started the implementation of the SPRA:

- EPRI 1025286 for short-term seismic walk downs, May 2012, (NTTF recommendation 2.3)
- EPRI 1025287, February 2013, called the Screening, Prioritization, and Implementation Details (SPID) for implementation of the full SPRA (NTTF recommendation 2.1)

What is the big picture, the major steps, of an SPRA?

The current NRC plan for SPRA is described in the EPRI SPID (EPRI report 1025287, 2013). The basic steps for US Nuclear Power Plants are diagramed in the EPRI SPID. These steps are, in broad terms:

- Steps 1 and 2: Develop new hazard curves and ground motion response spectra (GMRS) and foundation input response spectra (FIRS).
- Steps 3a through 3f: Perform a screening procedure based on ground motion spectra to determine which plants are required to perform new seismic risk evaluations
- Step 5: Determine whether a seismic margins assessment (SMA) is sufficient, or whether an SPRA is required.

- Steps 6 through 8: Perform the SMA or SPRA as appropriate and submit the results to the NRC.

But the fukushima accident was caused by the Tsunami, so why focus on the magnitude of the ground motion?

The potential and probability of a tsunami are also investigated for plants on the ocean shore, and seismic-induced failures of dams and external flooding are also considered for all plants. But an important issue addressed in the SPRA is the consequence of an extended station blackout: loss of off-site power and loss of EDG power, for a long period of time, causing loss of core decay heat removal capability, as happened at Fukushima.

In this context, the SPRA Project provides several safety benefits to plant operation:

- It permits to understand postearthquake accident vulnerabilities, and the expected role of control room operators in case of large earthquake or extended station blackout.
- It explains systematically the likelihood of postearthquake and post-blackout core damage, and the dominant risk contributors.
- It quantifies the risk to the core and risk to the population, and permits a comparison to other risks.

The objective of this effort is to pinpoint potential upgrades that would provide a better chance of accident prevention in case of station blackout or failures caused by a very large earthquake, larger than the design basis SSE. These upgrades could then be conceptualized for a cost–benefit analysis.

What is flexible coping strategies and how does it fit in the SPRA?

Early on in planning for the SPRA it became clear that a full SPRA project would take time. The plants and the NRC wanted an effective means of implementing more immediate improvements, while the SPRA proceeded. These improvements took the form of a new capability, or flexibility, to respond to BDBEs. This is the objective of FLEX: Diverse and Flexible Coping Strategies.

The objective of the FLEX strategy, coordinated through NEI, and submitted to the NRC, is to enhance nuclear reactor operators' ability to cope with severe natural phenomena hazards, such as those that occurred during the Fukushima Daiichi accident: extended station blackout caused by loss of off-site power and loss of on-site emergency generators, causing in turn a loss of RHR pumping capability, and loss of control room functions.

The FLEX strategy provides the means to respond to an extended station blackout by deploying and connecting backup emergency equipment, power supplies, and instrumentation and controls to maintain reactor core and spent fuel pool cooling for an extended period of time, and to ensure the integrity of the containment building should normal and backup systems fail.

The applicable references are:

- NRC Bulletin 2011–01: Mitigating Strategies, May 11, 2011
- Order EA-12-049: Issuance of Order to Modify Licenses with Regard to requirements for Mitigation Strategies for Beyond-Design-Basis External Events
- NEI 12-02: Industry Guidance for Compliance with NRC Order EA-12-051, Revision 01
- NEI 12-06: Diverse and Flexible Coping Strategies (FLEX) Implementation Guide
- Order EA-12-051: Issuance of Order to Modify Licenses with Regard to Reliable Spent Fuel Pool Instrumentation

One of the early steps in the SPRA is the development of the site seismic hazard, how is this done?

The objective of the probabilistic seismic hazard analysis is to develop the site-specific ground motion response spectra (GMRS). This activity can be subdivided into three phases:

- The seismic source model (seismicity of the region around the plant)
- The attenuation model (how the seismic waves propagate from the source to the site base rock)
- The site response (how the base rock motions at the site propagate up through the site rock and soil to the surface) to result in the GMRS

The GMRS are first used to determine whether an SPRA is necessary, following the EPRI SPID. If an SPRA is required, the GMRS will be used as input to the building seismic probabilistic analysis and to develop the in-structure response spectra (ISRS) for the evaluation of systems and components. An additional geotechnical task is to check the risk of geological hazard in the form of seismic-induced fault displacement, slope failure, liquefaction, and settlement.

The next step is the development of the system analysis logic, the event trees and fault trees?

The objective of the SPRA system analysis is to develop plant and systems PRA models, in the form of event trees and fault trees, to study seismic-induced accident sequences. Then, the fragility input is integrated into the models to quantify seismic-induced accident sequences. The results of the system analysis will be used to estimate core damage frequency (CDF) and the consequent potential for large early-release frequency (LERF). The system analysis must be based on the current as-built condition of the plant, and will therefore involve a system analysis walk down to ascertain the as-built condition of SSCs.

Then comes the SSCs fragilities?

Yes, a fragility is the conditional probability of failure of an SSC to perform its seismic function at a certain seismic input. Fragilities are developed for SSCs identified by the system analysis. The fragilities will be entered in the system analysis to identify seismic contributors to risk. Seismic fragilities for the building structures may rely on a finite element model of the building, coupled with an SSI of the building and soil layers. The same model is then used to develop ISRS to be used for the systems and components evaluations. The methodology for calculating fragilities is described in several publications: EPRI NP-6041 (1991), EPRI TR-103959 (1994), EPRI 1002988 (2002), EPRI 1019200 (2009), and NUREG/CR-5270 (1989), NUREG/CR-1150 (1990).

Fragilities are expressed in the form of high confidence of low probability of failure (HCLPF), together with the median capacity (A_m), the randomness variability (β_r), and the uncertainty (β_u), in terms of the peak ground acceleration, or the peak spectral acceleration. The data necessary for developing the fragilities are the existing seismic design analyses and test reports and calculations, combined with walk downs to ascertain the as-built condition of the plant. In addition, generic fragility values are used, and SSCs with a high HCLPF can be screened out.

What is the bottom-line measure of the plant's ability to sustain the beyond-design-basis earthquake?

There are two bottom-line numbers: the CDF and the LERF. The CDF is the probability of core damage, up to the potential for core melt. The LERF is the probability of loss of containment following core damage, and release and dispersion of radioactivity. The CDF and the LERF are calculated as output to the SPRA. In addition, the dominant contributors to the CDF and the LERF are identified, which permits the plant and the regulator to focus on their improvements.

Any thoughts on the benefits of SPRA?

The SPRA is a well-structured quantitative approach to address a difficult question: what would happen in case of a very large (beyond-design basis) earthquake? The SPRA permits us to identify vulnerabilities to beyond-design-basis earthquakes or floods, and as importantly, vulnerabilities to extended station blackouts where the plant has lost its off-site and on-site sources of power for a long period of time. It permits to have in place emergency response equipment and procedures to provide core cooling, and protect essential SSCs. SPRA is another example of how the industry shares information, learns from small incidents and from large accidents to any one of the many nuclear power plants around the world, and develops actions to prevent recurrence.

Since seismology progresses as more earthquakes occur, is there a retrofit impact on a plant's original OBE and SSE seismic design?

The answer is yes. A classic example is one plant for which a new seismic fault was discovered late in the design process. The plant committed and completed a long-term seismic retrofit project to account for the effect of the new fault.

A second example is the current understanding that the ground motions for the central and eastern United States (the so-called CEUS plants, which are the majority of US plants) are different than the western US-based ground motions that were used to develop the generic seismic design regulation. It became evident from the earthquake data that the ground motions for the CEUS plants have more high-frequency content, which would cause the GMRS to peak at more than 30 Hz instead of the western-based peaks at 5–10 Hz. The resolution of this CEUS GMRS issue was folded into the post-Fukushima SPRA project described above.

2.4 VIBRATION TESTING AND MONITORING

What are the causes of vibration in nuclear power plant systems and components?

The causes of vibration in nuclear power plants are not different than vibration in other industrial facilities: they can be mechanically induced or flow induced.

What is mechanically induced vibration?

Mechanically induced vibration of piping happens when the pipe is forced to follow and amplify the vibration of a piece of machinery to which it is attached at the nozzle. The mechanically induced vibration of the machinery itself (pump, fan, compressor, turbine) can be caused by several factors: axial or angular misalignment of shaft, loose anchorage to concrete (also referred to as soft foot), etc. These mechanically induced vibrations of machinery are well understood, and there are today well-developed tools and techniques to measure and diagnose such machinery vibrations, as part of routine maintenance and periodic testing programs under ASME O&M. The mechanical vibration of rotating equipment, and in some cases of a large flexible equipment shell, can cause in turn vibration of attached, more flexible, distribution systems. The vibration of the equipment itself, the vibration source, may be too small to be visible, but it can be amplified by the response of connected flexible piping, tubing, or duct systems.

What is flow-induced vibration (FIV)?

FIV is vibration is induced by the flow of fluid inside a pipe or duct, and is typically caused either by pressure pulsing or by turbulence.

- In the first case, FIV is caused by steady-state pressure pulsations on the discharge of pumps, compressors, or fans. The fundamental pressure pulsing frequency is the number of revolutions per minutes or strokes per minute, multiplied by the number of vanes, the vane pass frequency, the piston cycling frequency, or multiples (harmonics) of this fundamental frequency. The driving frequency can be observed as a fast Fourier transform (FFT) frequency spectrum. This is why, when facing FIV one of the most useful tools to understand its cause and therefore to solve it, is to measure the pressure pulse signal from the fluid inside the pipe or duct. This can be done through existing instrumentation or, more often, through purpose-installed high-frequency pressure transducers, connected to the main line through a vent or a drain.
- The second type of FIV is caused by flow turbulence, and in some cases cavitation, for example, downstream of valves, orifice plates, or sharp branch openings. It is a function of the flow velocity, and therefore may not appear at plant start-up, but will appear later when the flow conditions change, for example, increasing the flow rate.

Is resonance necessary to cause visible or damaging vibration?

No, resonance is not necessary to cause vibration, but of course it will amplify the vibration amplitude. FIV can be amplified by acoustic resonance and by structural resonance. In the case of acoustic resonance, the amplitude of the pressure pulse is amplified if the pulsating pressure frequency is close to one of the system's acoustic frequencies. In one case of acoustic resonance, pressure pulsations of approximately 350 psi have been measured in a piping system, oscillating around a mean system operating value of 2000 psi at a frequency near 300 Hz. These pressure pulses travel down the line and generate forces at flow discontinuities in the piping system (elbows, valves, reducers, tees, orifices, etc.).

In the case of structural resonance, the driving hydraulic frequency is close to one of the system's structural frequencies, so that the structural response is amplified. This is particularly true for FIV where the damping is small because the vibration amplitude is small. The natural structural frequencies can be those of a breathing mode (radial pulsing of the pipe or duct cross-section), or beam mode (bending of the pipe or duct run.)

What are the common forms of vibration in piping systems?

Vibration can be steady state or transient in nature. Steady-state vibrations are referred to as high cycle–low amplitude, and in the long run they can cause fatigue-induced cracking. Transient vibrations are often low cycle–high amplitude, they can cause large movements and loads on the piping system and its supports, and they are often associated with repeated water or steam hammer.

Why is FIV a concern?

Fatigue cracking of components and supports caused by FIV is one of the most common concerns for operating plant piping systems. In addition to fatigue cracking, there have been cases where FIV caused fretting wear. In one case, FIV downstream of an orifice plate caused surface wear marks on stainless-steel piping by abrasion against reflective metal insulation caps (IN 07-21). Even though the wear marks were shallow, only 1/36 inch deep, this is something to keep in mind. We are aware of a case, in a chemical plant, where similar wear, this time by contact between a vibrating pipe carrying hydrogen and a bolt head bearing against the pipe, almost penetrated the hydrogen pipe wall.

When is vibration monitoring necessary in nuclear power plants?

Vibration monitoring of piping systems in nuclear power plant is addressed in SRP Section 3.9.2. In practice, vibration monitoring is necessary in the following cases: (1) during the preoperational and start-up phase of a new plant; (2) after the installation of a new piping system, or after a significant layout or support modification; (3) during plant operation as part of the predictive maintenance of rotating equipment (pumps, compressors, fans); and (4) for diagnostics testing in case of an observed vibration anomaly.

What systems or components are monitored for vibration at the preoperational phase?

There are three classes of equipment that are vibration monitored at the preoperational stage: the reactor internals (start-up), certain piping systems (start-up and in case of anomaly), and rotating equipment (periodically).

What is the purpose of the piping preoperational vibration test?

Unlike heat exchanger tubes in cross-flow, piping is typically not designed for FIV because, as we said earlier, it is difficult to predict the hydraulic forcing functions that could cause the pipe to vibrate in service. There are a few exceptions, such as the design against acoustic resonance in a primary BWR steam system or PWR reactor coolant loop (RCL). Whether the system is designed or not designed against vibration, critical piping systems are monitored at start-up of a new plant to detect whether there is significant mechanically induced vibration or FIV in the system.

In a BWR, for example, nearly 50 subsystems are monitored either during preoperational testing or during start-up testing. Of these, approximately 40 subsystems are visually inspected for vibration, and 10 are instrumented for steady-state or transient vibration. In addition, small-bore piping and instrument tubing systems are typically visually observed for evidence of vibration.

What data is recorded in preoperational vibration testing of piping systems?

Vibration data recording consists typically of acceleration vs time, which is then integrated into velocity and displacement, and concurrent system pressure and flow vs time. The latter is important because it helps understand the hydraulic cause of the observed vibration.

Data gathering also includes the use of FFTs to transform parameters measured in the time domain into the frequency domain. The frequency domain data permits us to identify the dominant frequencies in a vibration signal, and deduce its cause.

What constitutes acceptable vibration in piping systems?

The criteria for vibration testing and monitoring are provided in the ASME O&M Code, Division 2 Part 3, simply referred to as O&M 3. The O&M 3 criteria are based on an endurance stress limit. If the vibration-induced stress in the piping is below the O&M 3 limit, it should not cause a fatigue failure in the long run. However, keep in mind that pipe stress is not the only criterion; vibration of piping may eventually fail in-line components (valves, instruments), pump or compressor components (such as shaft retaining bolts), or supports. Therefore, these effects of vibration on these components should also be evaluated, in addition to the stress limits on the pipe itself.

The allowable stresses are based on attributes such as configuration, material, weight distribution, concentrated mass, and stress discontinuities (captured in stress intensification factors (SIFs) and stress indices), and are obtained from measured vibration displacements, velocities, or accelerations.

What to do if a vibrating system does not pass the O&M 3 vibration limits?

The first thing to do is to determine the cause of vibration: Is it mechanically driven or is it hydraulically driven (FIV)?

Second, is to try to eliminate the source of vibration. If FIV, the pressure pulses may be eliminated by actions such as adding a pulsation damper to the system (either in-line or on a branch connection), by changing valve alignments, changing the number of vanes on a pump, or replacing valve internals by tortuous path internals to reduce vortex shedding.

Are these tortuous path valve internals always feasible?

This is something that engineering should pursue with the valve manufacturer. But be careful to check the risk of plugging because the tortuous paths have in many cases tight flow holes and channels. In one case the multistage tortuous path consisted of concentric cylinders with tight holes; they were installed inside a globe valve to accommodate a large pressure drop while preventing cavitation and excessive turbulence. But then, a postmaintenance test indicated a flow rate through this valve 10 gpm below the expected 75 gpm. The valve internals were inspected, indicating that several of the tight holes on the outer cylinder were partially plugged, and that these holes were smaller than the upstream strainer, so that debris could pass through the strainer and get caught in the tight orifices of the new tortuous path valve internals (IN 04-01).

What if the source of vibration cannot be eliminated?

In some cases it is not feasible to eliminate the source of vibration, or it may take time to implement the necessary hardware or operational changes. In this case, there are two parallel paths that should be followed:

- Analyze whether the vibration is acceptable, in other words is it a nuisance or can it lead to damage or failure. This can be done by following the stress or velocity evaluation procedures of ASME O&M 3.
- In parallel, investigate whether the vibration of the system can be dampened (adding vibration dampers, or adding shims between the pipe and its supports) or detuned (adding supports or rerouting the pipe). The net result is a change in mass, stiffness, or damping. Adding supports generally works in small-bore pipe. Adding dampers has proven to be a successful solution for the larger pipe.

FIV failures have been reported in socket welds to vents or drains; has the industry developed a solution?

Because fillet welds and threaded joints have proven to be prone to vibration-induced fatigue cracking, it is preferable to use butt welds, if practical, in systems subject to vibration. Alternatively, vibration testing of socket welds, sponsored by EPRI (EPRI Report TR-113890) revealed that socket welds with a ratio of length (along the pipe) to width (radial length of the filet) of approximately 2:1 have a much longer fatigue life when subject to high-cycle-low-stress vibration.

One plant experienced four socket weld failures in a 12-month period. The failed welds were downstream of pressure-reducing orifice isolation valves. The orifices themselves showed evidence of wear caused by erosion due to cavitation, with the orifice size growing by wear from 0.2 to 1 inch. This cavitation in turn caused FIV in the downstream piping (IN 98-45).

Which parameter (displacement, velocity, or acceleration) is best suited for the analysis of the effects of vibration in piping?

For continuously supported piping displacement can be imposed on a model of the piping system to obtain stresses that can be compared directly to the endurance limit. O&M 3 also provides an approximate criterion based on velocity. For cantilevered vents and drains, measuring acceleration at the valves yields accurate stresses at the branch connection. The O&M 3 criterion elaborates on these methods, the proper use of measuring vibration, and the level of analyses to be used, based on the complexity and magnitude of vibration.

How are preoperational tests for piping vibration Planned?

The plant piping systems are grouped into vibration monitoring groups (VMGs):

- VMG-1 is the group that will require detailed analyses to predict fatigue stress to be compared to the endurance limit. In addition to the critical systems (such as the RCL, the main steam, and feed water lines) VMG-1 may also include piping connected to pump suction and discharge lines, or connected to the steam condenser walls.
- VMG-2 will be evaluated using hand calculations following O&M 3 formulas, such as velocity criteria.
- VMG-3 is for less critical systems with low-vibration predictions; these can be visually screened during plant walk downs.

How are standby systems, those that do not flow during start-up or normal operation, monitored for vibration?

Standby systems that are not flowing during full power operation, are monitored for vibration during system surveillance flow tests.

Is vibration monitored beyond the initial plant start-up test?

Vibration monitoring is performed for new piping additions, modified system layouts, or changes that affect the hydraulic flow conditions, for example, an increase in flow rates, or the addition of a pump with different rotating frequency or a different vane design.

Which pumps and what parameters are tested in the preoperational phase?

Critical pumps such as the RHR pumps are instrumented and tested for vibration in the preoperational and start-up phases. For BWRs, other instrumented systems may include the core spray pumps, the high-pressure coolant injection (HPCI) pump, the reactor core isolation cooling pump, and the reactor recirculation pumps.

What is the purpose of in-service surveillance testing (IST) of rotating equipment?

Surveillance testing of pumps is addressed in ASME O&M Division 1 Subsection ISTB, O&M Division 2 Part 24 for the reactor coolant and recirculation pumps, and O&M Division 3 Part 14. The pump test program is typically captured in the plant FSAR Section 3.9.2. "Surveillance Testing of Pumps" which reports the monitoring of several parameters beyond vibration, such as pressure (head), flow, and thrust-bearing temperature.

Trending the data will help with the early detection of anomalies. Following repair and reassembly of pumps or motors, vibration testing is repeated before returning the pump to service.

How are reactor internals tested for vibration?

NRC RG 1.20 specifies the expectations for vibration testing of reactor internals. This test is also addressed in O&M 3 Division 3 Part 23, and is captured in FSAR Section 3.9.2. Because the steady-state and transient hydraulic forcing functions inside a reactor vessel are complex, vibration analysis of reactor internals depends to a great extent on measurements from a prototype reactor and from similar operating reactors. Finite element analysis (FEA) is used to predict the mode shapes of internals, and maximum permitted vibration amplitudes. These numerical results are validated by laboratory vibration tests on individual components of reactor internals and fuel assemblies.

At the preoperational stage, the reactor internals are subject to FIV under single-loop and multiple-loop flow conditions, for a minimum of 1 million cycles of vibration. Then, the vessel is drained and internals are closely inspected for evidence of fatigue or wear damage.

What are the examples of vibration-induced concerns that surfaced during plant operation, and how were they resolved?

Most vibration-induced issues are detected during preoperational testing; this is after all the objective of such testing. But vibration can also appear in service, particularly as a result of changes in flow rate, valve alignments, etc. The following experiences, presented in bullet form, describe instances of vibration-induced damage. Some of these examples have resulted in plant trips or system unavailability, and in a few cases longer-than-normal outages.

- Pump vane passing frequency: The failure of small-bore cantilevered connections, tubing, and pump fitting was caused by mechanical resonance of the pump base with the vane passing frequency. The pump impellers were reconfigured from four vanes to five vanes, and the vibration was reduced significantly. A flexible hose was added between the piping and tubing, and tieback supports were shimmed to zero gaps.
- Pipe supports attached to pump: The failure of a seal coolant pipe at the pump-threaded fitting was due to the fact that the piping was supported from the pump casing which amplified the casing vibrations. The piping support was modified to support the pipe from the pump pedestal base.

- Main steam instrument line at partial power: Cantilevered instrument lines on main steam piping failed by vibration. The vibration was observed to be most severe at partial power with the turbine control valves throttled. Once the plant reached full power the vibrations reduced significantly. The solution was to isolate overhang weight of the instrument lines using flexible braided hose. This reduced the vibration, but not enough. The flexible hoses were replaced with tubing which resolved the situation.
- Feed water instrument tubing: An instrument tubing line failed off a centrifugal feed water pump. This was a case of acoustic resonance. The instrument was not used and was not needed, it was removed.
- Filters acoustic resonance: The failure of threaded pipe in an oil piping was traced to acoustic resonance that had appeared with the addition of new in-line filters and piping. The filter-associated piping was modified and the acoustic resonance disappeared.
- Tieback of drain pots: High-amplitude vibrations were recorded on main steam drain pots that were cantilevered from the main line. Tieback supports were added which eliminated the problem.
- Failure of supports: When valves on small-bore warm-up lines were wide open, they caused two-phase flow mixing in the downstream header to the condenser, resulting in cavitation-induced vibration. This caused a spring rod to fail by fatigue at the turnbuckle and sagging of the line. When the small-bore valves on the warm-up line were throttled down to near-closed position vibrations on the downstream header disappeared.
- Cavitation downstream of orifice: A crack was found in a 6-inch elbow in a primary water pump recirculation line, downstream of an orifice. The crack initiated at the weld and propagated to the elbow base metal. Orifice-induced vortex shedding and cavitation appeared to be the root cause. When flow was diverted to a bypass line which did not have a flow-restricting orifice, the vibration ceased.
- Fretting wear: An air instrument line experienced external wall thinning at a support clip, caused by friction between the tubing and the support clip. The damage was due to fretting wear caused by vibration of the instrument line against the support clip.
- Flexible hose reduces vibration: A brass tube connected to a diesel-powered fire pumps failed because of turbulence downstream of a flow-restricting orifice. The tubing was relocated, and a flexible hose was installed to reduce vibrations.

- Resonance of a suspended switch: A suspended high-level switch was undergoing noticeable vibration. The cause of vibration was a resonance between the suspended switch and the main pipe. The switch was supported, eliminating the vibration.
- Anchor weld fatigue failure: The weld at an anchor between two valves failed as a result of FIV. In this case, the cause of vibration could not be readily eliminated. Instead, the weld size was increased to reduce the vibratory stresses, which prevented a repeated failure.
- Shell-mode vibration of condenser: A unit tripped due to a failure that occurred at a tube fitting, connected to a pipe which was in turn connected to the condenser. The source of vibration was traced to vibration of the condenser wall itself as the pressure in the condenser fluctuates. The tube route was modified to accommodate condenser wall movements and to minimize vibration levels.
- Tieback support on small line: A water leak was traced to a 1-inch branch line fillet weld to a 10-inch header. A tieback support was added, and proved sufficient to eliminate the vibration. A mechanical clamp was applied to contain the leak until a permanent repair was implemented.
- Rigid or constant load support: Two similar 36-inch steam lines were supported differently. In one case the line was supported by constant load hangers, in the other case the line was supported by a combination of spring hangers and rigid struts. The all-hanger layout caused nuisance noise, which was traced to vibration of internal parts of the constant load hangers.
- Expansion joint bellows fatigue failure: Damage occurred in the bellows of an expansion joint in the main turbine extraction steam piping. These expansion joints are designed to reduce nozzle loads on the main turbine. The cause of the bellows damage was traced to vibration-induced fatigue after nearly 20 years of high-pressure steam operation. The unit had to be shutdown to repair the bellows. Following the failure, the bellows were added to the Preventive Maintenance program for periodic replacement every 10 years.
- Resonant vibration: As a power plant ramps up from the shutdown condition to full power, the frequency of pressure pulsing inside certain systems also ramp up from zero to full values. In one case the driving hydraulic frequency was in resonance with pressure transmitter lines when the plant swept through the 10–30% power range, which at a power ramp rate of 3% per hour, lasted nearly 7 h, as the main steam

control valves are throttled. Over time, tubing with flexible hoses failed at the piping interface. The flexible hoses were replaced with lighter tubing similar to those used for another connection which did not experience visible vibration.

Can FIV damage valves?

Over the years, there have been instances of mechanical failures and separation of valve internals, and malfunction of internals. Some were due to workmanship issues with thread sizes, missing pins, internal welds and tack welds, and applied torque for thread engagement between parts; others were attributed to disassembly and reassembly procedures during maintenance; but many were caused by FIV. In one particular case, the antirotation pin in an MSIV sheared as a result of FIV; this caused the valve disk to drop into the flow stream, shutting down steam flow in the line, causing a reactor scram. Part of the repair consisted in adding a second pin, at 90° from the first one. Several information notices can be consulted on the subject of failures of valve internals: IN 81-08, IN 81-14, IN 81-28, IN 81-29, IN 81-30, IN 81-35, IN 82-20, IN 82-26, IN 82-35, IN 83-70, IN 84-33, IN 84-48, IN 85-67, IN 00-21, IN 06-15, IN 05-23.

2.5 BURIED PIPE

What types of pipes are buried in a nuclear power plant?

There are several piping systems that are buried in nuclear power plants, mostly in yards around the auxiliary and turbine buildings, and in some cases under the building foundation. These include water supply and return from the lake, river, or ocean; fire protection water lines; gas lines; or diesel fuel lines. Some are safety related and seismic category, others are not. Most are metallic, steel, corrosion-resistant alloys, or cast iron; they are bare, coated on the outside (with or without cathodic protection), or lined on the inside with cement or plastic. Some buried pipes are glass-fiber-reinforced plastic or high-density polyethylene.

How were buried lines in nuclear power plants analyzed and qualified in the original design?

The rules for design analysis of buried pipe have evolved in multiple directions for various industries: waterworks, oil and gas pipelines, and fossil power. But the rules for design analysis have not been addressed in the nuclear design code, ASME III. Instead, during the plant design phase, each architect–engineer developed project-specific procedures and standards.

Where to find more details regarding soil and surface loads?

In 2009, the ASCE published a "Manual of Practice" entitled *Buried Flexible Steel Pipe; Design and Structural Analysis*, which traces the history and development of buried pipe design, primarily for soil loads, and provides ample discussions on buried pipe design.

Professor A.P. Moser's *Buried Pipe Design* is another classic reference textbook for soil and surface load design. Watkins and Anderson published a comprehensive analytical treatment of soil and surface loads under multiple burial conditions entitled *Structural Mechanics of Buried Pipes* and another important description of soil and surface load effects is contained in Chapters 26 and 27 of Spangler and Handy *Soil Engineering*.

Were the soil and surface evaluation methods, described in these textbooks, documented in standards?

Yes, design for soil and surface loads is documented in AWWA standards such as AWWA M11 "Steel Water Pipe: A Guide for Design and Installation" and companion AWWA standards such as AWWA C150, and in "Welded Steel Pipe Design Manual, Merits, Design Standards, Technical Data and References," by the AISI.

Are the AWWA standards sufficient?

The AWWA standards focus on the effects of internal pressure, and soil and surface loads. As described in the Welding Research Council (WRC) Bulletin 425 "A review of Methods for the Analysis of Buried Pressure Piping" there are other loads that need to be addressed in certain systems, for example, constrained thermal expansion in hot buried pipe (such as buried steam lines), natural ground settlement, flooding, and seismic effects.

Were standards developed for loads other than AWWA for soil and surface loads?

Regarding constrained thermal expansion, the analysis method was developed by E.C. Goodling and documented in Appendix VII to ASME B31.1. This method accounts for the effect of soil stiffness in the form of pipe–soil spring interaction, and the effect of pipe–soil friction which results in a virtual anchor. A different approach which treats the soil as perfectly rigid is provided in the pipeline codes ASME B31.4 (liquids) and B31.8 (gas).

Regarding earthquakes, there are two effects to be addressed: A seismic wave passage effect, and a seismic anchor motion effect. The study of seismic response of pipelines has been of primary interest to two industries: The oil and gas pipeline industry and the nuclear power industry, and more recently, through the American Lifelines Alliance, to the waterworks industry.

Each industry developed standards and guidance for seismic analysis of buried pipes and pipelines. They are as follows:

- The oil and gas pipeline industry's ASCE Guidelines for the "Seismic Design of Oil and Gas Pipeline Systems," ASCE, 1984, with D.J. Nyman as the principal author
- The nuclear industry's ASCE 4 "Seismic Analysis of Safety-Related Nuclear Structures and Commentary," and ASCE "Manuals and Reports of Engineering Practice No.58, Structural Analysis and Design of Nuclear Plant Facilities," Section 7.6 Buried Pipe and Conduits, 1980, and "Seismic Response of Buried Pipes and Structural Components" which is a report by the Seismic Analysis Committee of the ASCE Nuclear Structures and Materials Committee, 1983

Is it not confusing to navigate all these references?

This question arose in the early 1990s. It is in fact a broader question: does buried pipe design belong to civil engineering, or to mechanical engineer discipline? In reality it is both, civil and mechanical, and we must add that it also involves corrosion engineering since so many buried pipe issues are traced back to soil-side or fluid-side corrosion. In the early 1990s, this multidiscipline aspect of the problem prompted ASME and ASCE to form a joint Task Group to compile buried pipe design guidance. The objective was not to create a new standard, but to help the designer navigate the many existing ones. This effort was supported by FEMA and resulted in the American Lifeline Alliance (ALA) report entitled "Guidelines for the Design of Buried Pipe." This report is a compilation of all loads applied to buried pipe, and an overview of the analytical formulas for their evaluation.

Are hand calculations sufficient to analyze buried pipe?

Most of the time, yes, but not in all cases finite element analysis (FEA) has an important role in two cases:

- First, FEA is useful when closed-form solutions are too conservative and too penalizing. This is the case with the analysis of constrained thermal expansion with the assumption of an infinitely rigid soil, as is done in the pipeline codes ASME B31.4 and ASME B31.8. In these cases a nonlinear soil pipe model, with soil springs and soil pipe breakout force, would result in more realistic strain and stress predictions.

- Second, FEA becomes necessary when buried pipes are subjected to large ground movements that plastically deform and may buckle the pipe or pipeline. This is the case, for example, for buried pipelines crossing active faults. It is not the case in nuclear power plants, where the soil under safety-related buried pipes is seismically stable, not prone to liquefaction or failure for earthquakes with magnitude up to the design basis SSE. But ground failure may be a concern for earthquakes larger than the design basis SSE, what is labeled a beyond-design-basis earthquake (BDBE). This is of interest to seismic probabilistic risk analysis (SPRA) of nuclear power plants under NRC SRP (NUREG-0800) Chapter 19, or SPRA reviews that resulted from the Fukushima accident.

While ASME III does not explicitly address buried pipe, ASME XI does, why such disconnect?

As plants grow older, understanding the corrosion condition of buried pipe becomes more urgent. The nuclear power industry, under the direction of EPRI, formed the Buried Pipe Integrity Group (BPIG). Among BPIG's priorities was the development and implementation of corrosion inspection programs. As plants inspect corroded buried pipe they need a criterion to determine whether the aged pipe is safe, fit, for continued service. Similar criteria exist for oil and gas pipelines in the form of ASME B31G whose focus is primarily internal pressure, while nuclear power plant buried pipes have to address all the loads: soil, surface, thermal, pressure, flooding, seismic, settlement, etc.

The need for criteria to evaluate the safety of corroded pipe in nuclear power plants resulted in the development of an ASME XI Code Case N-806. As a first step in the evaluation of corroded buried pipe, the plants may have to update their design basis calculations for the nominal (uncorroded) pipe, and document the design margin of the original nominal pipe, before the measured corrosion is evaluated.

In summary, what should the design engineer use today for design of buried pipe?

Until ASME III publishes methods and criteria for the analysis of safety-related buried pipes, and there is currently an effort underway to develop such a document in the form of a Code Case, the designer may refer to the various codes, such as ASME III for internal pressure, AWWA for soil and surface loads, ASME B31.1 Appendix VII for constrained thermal expansion,

ASCE for seismic wave passage and anchor motions, and ASCE for flooding. The ALA report "Guidelines for the Design of Buried Pipe" is a road map for metallic pipe; and WRC 425 is a reference for the technical basis of the various formulas, while ASCE "Seismic Response of Buried Pipes and Structural Components" is a good reference for seismic analysis of buried pipe. For corroded buried pipe, there is ASME XI Code Case N-806.

2.6 SPECIAL ANALYSES

2.6.1 Fatigue

What causes fatigue failure?

Fatigue failure is the formation and propagation of cracks due to a repetitive or cyclic load. Most fatigue failures are caused by cyclic loads significantly below the loads that would result in yielding of the material. The failure occurs due to the cyclic nature of the load which causes microscopic material imperfections (flaws) to grow into a macroscopic crack (initiation phase). The crack can then propagate to a critical size that results in structural or pressure boundary failure of the component.

Fatigue cracks normally initiate at stress concentrations, structural discontinuities. Fatigue cracks can also propagate from existing macroscopic cracks, such as weld defects. Fatigue cracking can be superimposed to a corrosion mechanism, and the combination of both effects (stress and corrosion) constitutes stress corrosion cracking (SCC).

What would be a good overall engineering evaluation approach to fatigue?

In a broad sense we could divide the question of fatigue damage in two general categories: high-cycle fatigue and low-cycle fatigue.

- In the case of high-cycle fatigue the stress amplitude is small but the number of cycles is so large that the designer must either keep the resulting stresses below the material endurance limit, or eliminate the cause of high cycles. High-cycle fatigue is typical of FIV or thermal cycling.
- Low-cycle fatigue consists of events causing large swings of stress. Low-cycle fatigue is typically caused by plant transients, large pressure, or temperature fluctuations from unique events, such as rapid heat up or cool down, or safety injection of cold water into a hot system. Unlike high-cycle fatigue, low-cycle fatigue events can be defined individually, and can therefore be analyzed and accounted for in the design life of the component.

What are the techniques used to analyze components for fatigue?

There are two parts to fatigue analysis: (1) the prediction of the cyclic load (the demand), and (2) the evaluation of the calculated cyclic stresses compared to allowable limits, including safety margins (the fatigue capacity). The prediction of cyclic loads is not a simple task. Cyclic loads can be due to FIV, local thermal gradients, expansion, or contraction.

The calculation and analysis of cyclic stresses is explicitly addressed in ASME III Class 1 systems and components. ASME III Class 1 components are designed by analysis to limit the cyclic stresses caused by thermal transients and thermal gradients, and by five postulated operating basis earthquake (OBE) which are Service Level B events.

There are three general techniques that have been and continue to be used to study the significance of cyclic stresses.

- Component tests: The first fatigue evaluation technique, and the older one, is based on cyclic tests of full-scale components. Piping component tests were first conducted by A.R.C. Markl and his colleagues in the late 1940s and early 1950s to develop stress intensification factors (SIFs) for piping components such as tees and elbows. The SIF approach is still applied in the analysis of ASME B31.1 and ASME III Class 2 and 3 piping systems.
- Polished bar specimen tests: The second fatigue evaluation technique is based on tests of polished bar specimens, from which stress–cycle (S–N) curves are developed for several materials. An adjustment factor of 2 on stress and 20 on cycles is applied to the failure S–N curves to obtain the ASME III design fatigue curves. The fatigue curves are published in ASME III Appendix I. The calculated stress in a component is first penalized by a stress concentration factor to reflect the stress concentrations caused by shape discontinuities and local plasticity in the component. Recent experimental work sponsored by the USNRC and by industry has resulted in revisions to the fatigue margins in ASME III Appendix I to account for the primary water environment (in particular boric acid, temperature, and oxygen concentration) as well as the strain rate of the cyclic load. The fatigue corrections that account for these primary water environmental corrections are resulting in fatigue limits that are referred to as environmental fatigue, which will be discussed later.
- Fracture mechanics tests: The third fatigue evaluation technique is based on fracture mechanics. The fracture mechanics approach studies the propagation of an initial crack. Fracture-mechanics-based fatigue analysis is the technique addressed in ASME XI, and is the subject of current

studies within ASME III to develop a flaw-tolerant design approach. The fracture analyses involve computing a critical crack size or the remaining life of a component subjected to cyclic or steady-state stresses. Another approach to fracture mechanics is probabilistic fracture mechanics which provides a probability estimate of fatigue life by considering the randomness of the applied loads and the variables.

What are the ASME III fatigue curves?

The ASME III design fatigue curves are developed experimentally and used to qualify Class 1 nuclear plant components for an original plant design life of 40 years, or an extended life of 60 years, and possibly beyond.

The ASME III fatigue curves plot the allowable number of cycles N as a function of the applied, elastically calculated alternating stress intensity S, for different materials, the (S,N) curve. The elastically calculated stress S entered in the (S,N) fatigue curve is an artifice, a simplification to represent fatigue test results which are in the form of imposed plastic strains ε versus cycles to failure $N_{failure}$. The equivalent elastic stress S is then obtained as $S = \varepsilon E$, the strain multiplied by the modulus of elasticity, with corrections for plasticity effects. This explains why these fatigue curves have ordinates in the order of hundreds of kilopounds per square inch(ksi) when the material ultimate strength is not higher than say 60 ksi.

Are experimental fatigue failures defined as the fracture of the test specimen in two?

There are several ways to declare failure in fatigue tests.

- In the first definition, cycles to failure $N_{failure}$ are counted as the number of cycles to cause fracture of a test specimen in two halves, but to be valid such a definition must be based on a standard specimen size.
- A second definition is to define the cycles to failure $N_{failure}$ as the number of cycles that result in a reduction by half of the stiffness of the specimen, as it is subjected to reversing strain of a constant amplitude.
- A third definition of $N_{failure}$ is the number of cycles at which the applied load (magnitude of the cyclic force) to achieve a certain strain (elongation) drops by say 50% or by 25%. The last load drop, by 25%, happens to correspond to a small crack, about 3 mm long, something that can be detected by surface nondestructive examination methods, and can be labeled an engineering crack (detectable) rather than a microcrack which would be in the order of 300 μm. The 25% load drop definition of failure has been used in practice to define failure when performing polished bar fatigue tests.

What are the characteristics of a fatigue crack compared to an overload fracture surface?

Close examination of fatigued specimen reveals two distinct regions. One is smooth, a region of slow crack growth with concentric beach marks around the location of the initial flaw. The beach marks occur when the crack is arrested before resuming its growth. Eventually, the crack would become large enough to satisfy the energy or stress intensity criteria for rapid propagation. This is the second region where sudden crack propagation occurs marked by high surface roughness due to tearing. It is normally possible to correlate the beach marks with stress cycles, and to estimate the applied stress at failure from the size of the crack just before rapid propagation and from the fracture toughness of the material.

How were the ASME fatigue design curves developed?

The ASME fatigue design curves were developed experimentally, for various materials. Starting with the best fit of strain–cycle failure results, the data is adjusted for mean stress effects, and then the S–N curves are shifted by a factor of 2 on stress, and a factor of 20 on cycles. In other words, if a fatigue failure occurred in 20,000 cycles during testing, the ASME design fatigue curve would permit no more than 1000 cycles of operation at that stress.

Originally, the adjustment factor of 20 on cycles was meant to account (1) material variability (a contributing factor of 2), (2) size and homogeneity differences between the polished bar test specimen and an actual fabricated component (a contributing factor of 2.5), and (3) surface finish differences between the polished bar specimen and the actual plant component (a contributing factor of 4, an important and much studied effect). Thus, multiplying the three contributing factors we obtain the factor of $2 \times 2.5 \times 4 = 20$, which was the factor applied to the experimental number of cycles.

Do the factors of 2 and 20 built into the ASME III design fatigue curves account for the reactor water environment?

The factors $2 \times 2.5 \times 4 = 20$ do not explicitly account for a corrosive environment. But the factors of 2 and 20 were judged by some to be sufficient to account for the corrosivity of the water and steam inside pipes and equipment. It is this point—whether the factors of 2 and 20 are sufficient to account for the corrosivity of the environment on fatigue life—that was questioned, starting in the 1980s, and which has evolved to become what is labeled today the environmentally assisted fatigue (EAF) concern, which we will discuss later.

What is the role of the mean stress in fatigue analysis?

Fatigue loading is cyclic in nature, from σ_{max} to σ_{min} around a mean stress σ_m. If the mean stress is zero where the load goes from fully compressive to fully tensile at a given point along the member, then the load is known as fully reversed fatigue load. The ASME III fatigue curves are corrected to account for a large mean stress, equal to yield of the material, such as would be caused by residual stresses in welds. Another parameter is the stress ratio $R = \sigma_{min}/\sigma_{max}$. For example, a value $R = 0.1$ represents a tension–tension cycle where the minimum cyclic stress $\sigma_{min} = 0.1\sigma_{max}$ or 10% of the peak stress. For a fully reversed loading, $R = -1$.

Obtaining an (S,N) fatigue curve is not a small task even for the simplest of cases where the load is fully reversed. Significant amount of testing would be needed, as well as different combinations of mean and alternating stress. To overcome these difficulties, methods were developed to enable the estimation of the endurance limit for a given set of mean and alternating stresses; the most commonly used is the Goodman diagram.

How are the ASME III fatigue curves used in design?

We can describe the ASME III Class 1 fatigue analysis procedure in steps:

- The first step of a fatigue analysis is to define the cyclic loads and transients that need to be analyzed. This is done in the ASME III Design Specification.
- The transients are applied to a heat transfer model of the fluid and the metal component, and the heat transfer analysis is used to calculate the cyclic stresses caused by each of the thermal transients. Because fatigue tends to be governed by local stress concentrations, the calculated stresses must include local peak stress effects. The total stresses, including the peak stresses, are obtained either by FEA as is the case in ASME III NB-3200, or by applying stress indices to the nominal stresses, as is the case for piping in ASME III NB-3600.
- The designer then verifies that the cyclic stress will not ratchet, which means that the strain, even if plastic, will not accumulate at each cycle. This ratcheting check is performed following an ASME III formula with a $3S_m$ stress limit.
- The maximum range of the stress intensity (twice the maximum shear stress) is calculated for each transient, and the alternating stress intensity

S_{alt} is half the largest range of stress intensity. A plasticity correction factor K_e is applied if the stress is large. The alternating stress intensity $(S_{alt})_I$ is calculated for each Class 1 component and each transient.

- Entering $(S_{alt})_i$ into the ASME III Appendix I fatigue curve we read the corresponding number of allowable cycles $N_{allowable\text{-}i}$. The usage factor is then defined as the number $N_{actual\text{-}i}$ of actual cycles at $S_{alt\text{-}i}$ that will accumulate during the plant life, divided by the allowable cycles $N_{allowable\text{-}i}$ obtained from the fatigue curve: $U_i = N_{actual\text{-}i}/N_{allowale\text{-}i}$.
- Repeating this procedure for each transient (i) and adding all the usage factors, we obtain the total usage factor U at a given location as the sum of all the U_i at that location, $U = \Sigma U_i$. ASME III puts a design limit $U < 1.0$ to qualify Class 1 components.

Do these procedures apply only to ASME III class 1 components?

Yes, according to ASME III these explicit fatigue usage factor analysis procedures apply to Class 1 components governed by Subsection NB, not to Classes 2 and 3 governed by Subsections NC and ND, and they are not required in nonsafety ASME B31.1 components. However, if a cyclic fatigue concern arises for non-Class 1 components, the Class 1 procedure can provide a template for analysis.

How is fracture mechanics applied in fatigue analysis?

The classic usage factor approach to fatigue $U = \Sigma U_i < 1.0$ is meant to protect against fatigue crack initiation and propagation. In many cases, most of the fatigue life is spent on crack initiation and once that occurs, fatigue crack propagation can be rapid. However, performing crack propagation analyses is essential in evaluating the potential for failure of a component with an existing crack or a cracklike defect. Experiments have shown that the crack growth rate per cycle da/dN correlates with the cyclic variation in the stress intensity factor ΔK, in the form $da/dN = A \cdot \Delta K^m$, where A and m are parameters that depend on the material, the environment, the frequency of the load, the temperature, and the stress ratio. This relationship $da/dN = A\, \Delta K^m$ is known as the Paris law.

In the case of corrosion-induced cracking, without cyclic fatigue, the relationship between the crack growth with time da/dt and the applied stress intensity K would ideally be reflected by the relationship $da/dt = B \cdot K^n$ but the experimental correlations tend to have more variability than the da/dN predictions. In 1968, J.R. Rice introduced the concept of the "J" integral to fracture mechanics, which is used to achieve better predictions than by linear elastic analysis using the stress intensity K.

How did the EAF concern and its resolution evolve over the years?

The question of environmental assisted fatigue (EAF) as opposed to fatigue in air (as is the case for the (S,N) curve of ASME III Appendix I) was raised starting in the 1980s. In 1995, the matter was elevated by the NRC to a Generic Safety Issue (GSI)-166. Then, in 1999, a new NRC GSI-190 required that the question of environmental fatigue be addressed for the assessment of plant life extension, from the original 40 years to an extended life of 60 years, under 10 CFR 40.41 Aging Management Program. That same year ASME was requested by the NRC to address how environmental effects affect the ASME III Appendix I design fatigue curves. The question was studied within the PVRC, and results published as WRC Bulletin 487 in 2003, in what had become an international effort.

The NRC, working with Argonne National Laboratory (ANL) with significant international contributions, compiled fatigue test data in reactor environments, and conducted additional tests. Currently, these studies have compiled nearly 1000 data points for ferritic steels, nearly 600 data points for stainless steel, and nearly 150 data points for nickel alloys.

The work by the NRC and ANL resulted in the publication of Regulatory Guide 1.207, drafted in 2006, and issued as final in 2007, with the supporting data presented in NUREG/CR-6909, "Effect of LWR Coolant Environments on Fatigue Life of Reactor Materials," 2007. These two documents, RG 1.207 and NUREG/CR-6909, became the subject of much review by the nuclear engineering community, and a revision to NUREG/CR-6909 is in progress at the time of this writing.

What is the current ASME approach in resolving the question of EAF?

The ASME committee that had been studying the matter, and had reviewed in detail the NRC–ANL studies, published two Code Cases to address EAF: CC N-761 and CC N-792.

- The first Code Case (N-761) addresses EAF by revising the fatigue curves, reducing the number of allowable cycles for a given stress when compared to the current non-EAF curves. It has not been approved by the NRC, for several reasons.
- The second Code Case (N-792) addresses EAF by applying an environmental fatigue factor F_{en}, a penalty factor, to the calculated usage factors, where F_{en} is the ratio of allowable cycles $N_{air}/N_{environment}$.

An alternative approach to EAF is being investigated using a fracture mechanics approach referred to as flaw-tolerant design. In the flaw-tolerant approach an initial crack is postulated and then its growth is analyzed when subjected to plant transients in the primary water environment.

What does the penalty factor F_{en} depend on?

The studies described above indicate that F_{en} is a function of the type of metal; the sulfur content in the metal (weight percent); the metal temperature (degrees); the oxygen content in the water (parts per million); the strain rate of the cyclic load, i.e., how fast the metal is strained during the fatigue cycle (1/s); and the applied strain amplitude. A Code Case is in development to address the calculation of strain rate needed to calculate the F_{en}.

What design measures should be considered in a structure or component subject to fatigue damage?

Ideally, we would want to eliminate the source of cyclic loading. But in many cases this is not feasible without affecting plant operation. If we cannot prevent the cyclic loading, we have to mitigate its effects or reduce its magnitude, such as reducing points of stress concentration. Figure 2.3 is a case in point, where the pipe is connected by threaded joints. In

Figure 2.3 Piping with threaded pipe joints.

Figure 2.4 Similar piping with butt welded joints.

a similar system, Figure 2.4, the threaded joints were replaced by butt welded joints to eliminate the source of possible fatigue cracking.

In some cases, fatigue was due to thermal cycling caused by temperature transients or thermal mixing of two streams at different temperatures. In this case thermal sleeves have been used to reduce the thermal cycling stress in the nozzle or tee.

What has been the track record of thermal sleeves?

Thermal sleeves are cylinders welded inside pipe nozzles or mixing tees to reduce the thermal gradient effects on the pipe wall. In some of the early applications thermal sleeve welds failed after a while in service. The fractured sleeves either stayed trapped in position (IN 82-09), or they were carried downstream by the flow. This was the case for a 10-inch thermal sleeve that was initially installed in the accumulator injection nozzle to the cold leg of the RCL (IN 82-30). Radiography can be used to confirm the presence (or failure) of thermal sleeves. The cause of thermal sleeve failures have been FIV or thermal cycle fatigue (IN 82-09, IN 82-30).

2.6.2 Thermal Stratification

What is thermal stratification, and where was it observed?

Thermal stratification is the layered flow of slow-moving fluid. Thermal stratification occurs in certain piping systems in nuclear power plants. Thermal stratification was first recognized in 1979 in the feed water lines. Then, the stratification of flow in the pressurizer surge line was noticed in the 1980s in the United States. It was measured shortly afterward at other PWRs. What causes the flow to stratify in the horizontal legs of a surge line is the difference in density between the hot fluid surging out of the pressurizer and the relatively colder fluid nearly stagnant in the hot leg and in the surge line pipe.

Another instance of thermal stratification occurs if an isolation valve leaks slowly from a high-pressure side into a low-pressure side at a different temperature. Because leakage occurs at low flow rate, the colder and hotter liquids tend to stratify rather than mix, with the hot water layer flowing on top of the colder heavier layer.

Why is it important to understand which lines can stratify?

Stratification can cause two possible problems: bowing and striping. Because the top of the pipe metal is at a hotter temperature than the bottom of the pipe, and if the gradient of temperature is sufficiently large, the pipe will tend to bow upward. The second effect, striping, is the temperature fluctuation that occurs at the hot-to-cold fluid interface. Because the change of temperature, top to bottom, is gradual there is not a sudden step change of temperature, and striping has proven to be of little practical consequence.

How to predict the thermal stratification temperature profiles?

First, based on operating experience and on a systematic review of piping layouts, the locations with potential stratified flow are identified, including valve-leakage-induced stratification. These lines are then identified in the Design Specification as having a stratification load case. The operating conditions leading to stratified flow, with their number of cycles, should also be defined in the Design Specification.

For the locations of stratified flow, the boundary conditions (the hot end and the colder end), and the flow regime in the line are defined. Unless measurements are available for the full spectrum of operating conditions, a CFD analysis is conducted to determine the flow structures in the line (mixing and stratified regions) and the stratified temperature patterns and gradients in the line, as a function of time, for several transients. The CFD model is then benchmarked against published, measured stratification data. Typically several flow conditions are analyzed before the worst case, largest thermal gradient, can be determined. The CFD model mesh must be sufficiently detailed to capture global stratification and local striping effects. The fluid model will include temperature-dependent buoyancy, density, viscosity, specific heat, and conductivity. The model should include turbulent effects to reflect the flow mixing at the mixing nozzles.

How to analyze thermal stratification stresses?

The stress analysis of the stratification transient can be done either by stress linearization techniques of ASME III NB-3200, or by the pipe stress analysis approach of ASME III NB-3653.

In the stress analysis, three thermal-induced stress effects should be considered:

- The beam bending effect: The stratified temperature profile causes a beam bending effect of the pipe, and moments along the pipe.
- The transient effect: During the time that stratification is initiated or is terminated, the pipe wall will be subjected to through-wall temperature fluctuations. These effects contribute to the ΔT_1 and ΔT_2 and $T_a - T_b$ terms in the NB-3650 analysis.
- The steady-state effect: During steady-state stratified flow conditions, the temperature gradient from the top of the pipe to its bottom, at any cross-section, can be divided into a constant, uniform, temperature, plus a linear temperature variation from top to bottom, plus the remaining nonlinear top to bottom temperature profile, with a maximum value ΔT_3 of the nonlinear profile, which causes $E \alpha \Delta T_3$ stresses. This too is where the striping stress belongs, as an additional ΔT_3 effect.

How is thermal stratification actually monitored in the plant?

Temperature sensors can be placed at various monitoring locations along the piping. Resistance temperature detectors, are attached to the pipe outer surface, corrected for heat transfer through the pipe, and correlated to the plant data.

What are useful references related to stratification?

The analysis of understanding, assessment, and analysis of thermal stratification in nuclear power plants is addressed in several technical papers and the following references: NRC Bulletin 88-08 (stratification in lines connected to the reactor coolant system), NRC Bulletin 88-11 (stratification in the surge line), NRC Bulletin 89-80 (stratification in the HPCI line), and EPRI Material Reliability Program (MRP) reports MRP-146, MRP-132, MRP-170, and MRP-192.

2.6.3 Pipe Rupture

What is the overarching principle of pipe rupture hazards analysis (PRHA)?

The word rupture is used to describe breaks or cracks. There are two kinds of breaks: circumferential (guillotine) and longitudinal. There are two types of cracks: through-wall cracks that are postulated (non-mechanistic), and leakage cracks that are based on leak-before-break (LBB) analysis.

Ruptures have to be postulated in piping systems, and the effects of these ruptures must be mitigated. Because nuclear power plants are designed, operated, and inspected to prevent such ruptures, these are postulated events, they are sometimes referred to as non-mechanistic to emphasize that they are postulated despite the fact that the piping is designed and inspected to prevent ruptures.

What governs the postulation of pipe ruptures?

Pipe rupture hazards analysis (PRHA) is the set of analyses and qualifications to address 10 CFR Part 50 Appendix A, GDC 4, Environmental and Dynamic Effects Design Bases, which states:

"Structures, systems, or components important to safety shall be designed to accommodate the effects of and to be compatible with the environmental conditions associated with normal operation, maintenance, testing, and postulated accidents, including loss-of-coolant accidents. These structures, systems, or components shall be appropriately protected against dynamic effects, including the effects of missiles, pipe whipping, and discharging fluids, that may result from equipment failures and from events and conditions outside the nuclear power unit. However, dynamic effects associated with postulated pipe ruptures in nuclear power units may be excluded from the design basis when analyses reviewed and approved by the Commission demonstrate that the probability of fluid system piping rupture is extremely low under conditions consistent with the design basis for the piping."

The idea is to postulate ruptures (breaks and cracks) in pipes, investigate the effects of these breaks on the plant SSCs, and make sure that the plant can be safely shutdown and remain in safe shutdown if one of these ruptures were to occur. The SSCs relied upon to achieve this safe shutdown are called essential.

Are essential SSCs the same as safety class SSCs?

No, essential SSCs are those SSCs needed to safely shutdown the plant and keep it in a safe shutdown mode given a particular initiating break. In other words, an SSC is essential or nonessential in reference to a particular postulated break.

How is PRHA linked to LBB?

The last sentence of GDC 4, quoted above, the one which starts with "However," recognizes that a pipe that qualifies for LBB will not suddenly break; it will leak, and the leak will be detected and mitigated before it breaks. For pipes where this LBB behavior can be established, dynamic effects from breaks need not be postulated. But there are three important cautions, regarding LBB:

- First is that LBB does not eliminate the analysis of flooding effects and environmental effects from breaks.
- Second is that LBB does not eliminate the sizing of the emergency core cooling system (ECCS) system for a break.
- Third is that LBB does not eliminate the need to consider the effects of pipe whip and jet impingement on the blowout of insulation and the consequent clogging of the containment sump, which is treated as an ECCS function rather than a dynamic effect. This third effect is addressed in NEI 04-07 and GL 2004-02.

What is meant by PRHA?

The word "rupture" refers to the postulated failures: breaks (circumferential and longitudinal breaks) and cracks (through-wall and LBB-based leakage cracks.)

The "hazards" refer to three types of effects: dynamic effects, environmental effects, and flooding effects. Flooding can be folded under the environmental label.

- Breaks are postulated to occur in high energy (HE) lines (operating temperature in excess of 200 °F or operating pressure in excess of 275 psig) and, in some cases, in moderate energy (ME) lines. Breaks are either circumferential or longitudinal. In turn, circumferential breaks (commonly known as guillotine breaks) are either full separation if the thrust force is sufficient to fully separate the two ends of break (which is commonly the case), or the circumferential break may be of limited separation if the pipe is so rigidly restrained that it cannot move sideways more than its thickness (which is rarely the case). In some of the older plants the definition of HE has an "and" in place of an "or," so that HE in these older plants are lines operating above 200 °F and 275 psig.

- Through-wall cracks are postulated to occur anywhere around the circumference of HE and moderate-energy lines.
- Leakage cracks are postulated cracks that have a deterministic nature because their location and size are based on an LBB fracture mechanics analysis and on the leak detection capability inside containment.

Where are the methods and criteria for PRHA spelled out?

The methods and criteria for PRHA are described in SRP Sections 3.6.1 and 3.6.2, and ANS 58.2. While the SRP sections have remained unchanged since the 1970s, some aspects of ANS 58.2, such as the shape of the characteristics of the jet emanating from a break, have evolved in the 2000s, and documented in NEI 04-07 in response to GL 2004-02.

Where does PRHA fit in the overall design qualification of a nuclear power plant?

PRHA has an important role because it leads to large loads, often the largest loads, larger than seismic loads, on the plant structures. It also can result in a change in plant layout to prevent a whip or jet interaction on an essential target. PRHA will result in the addition of whip restraints and, in some cases, jet shields which are large structures. PRHA also governs the ambient temperatures and pressures for environmental qualification. PRHA must be one of the early layout and design considerations, and not be left for the later stages of design.

What should be the key elements of an analysis plan for PRHA?

- First, is the delineation of scope boundaries on P&IDs, in particular three types of boundaries: the boundaries of HE lines, the boundaries of break exclusion zones (BEZs) located between containment isolation valves, and the boundaries of the LBB lines.
- Second, within the boundaries of HE lines, the project must define and label each break.
- Third, the thermohydraulic parameters of the line must be established, both at power prior to the break, and after the break as the pipe blows down its contents. This information will be needed to calculate the dynamic, environmental, and flooding effects of the postulated ruptures.
- Fourth is the analysis of the four dynamic effects associated with breaks: (1) pipe whip from thrust forces at the break, (2) jet impingement on

nearby targets, (3) blowdown water hammer in the broken line, and (4) mass and energy releases leading to the prediction of subcompartment pressures and temperatures.

- Fifth, the HE and ME breaks and through-wall cracks are compiled and grouped by room or zone, to select those that will lead to the bounding flood levels and environmental effects.
- Sixth, having established the demands caused by the ruptures (dynamic, environmental, and flood), the project determines how these loads affect essential SSCs. Where essential SSCs can be adversely affected, preventive measures must be implemented: moving the essential target, qualifying the target for the load, restraining the broken pipe, or installing jet shields and whip barriers.

What measures are used to counter the effects of pipe ruptures?

The preferred method is to prevent the adverse interaction by relocating the essential targets. Where this is not feasible, then pipe whip restraints (PWRs), bumpers, jet shields, and barriers are used to limit the effects of ruptures. Also, the target can be designed and qualified to operate in the environment caused by the rupture.

How are whip restraints designed?

Whip restraints, jet shields, and barriers are designed using conservative energy methods or by dynamic nonlinear FEA, such as illustrated in Figure 2.5. There are two general categories of PWRs, rigid and plastic.

- Rigid PWRs are steel frames that are designed to elastically resist the whip force. They are typically avoided because they provide no cushioning effect to the whipping pipe, and as a result they transmit the full force to the backup steel and concrete structures. However, they are used to limit piping deflection in a congested area where we may want to minimize the zone of influence (ZOI) of the break.
- Plastic whip restraints are designed to plastically absorb as much of the pipe whip energy as possible. Examples of plastic PWRs are bumper pipes (Figure 2.6), U-bars (Figure 2.7), and honeycomb crushable pads. They have stress, energy absorption, or plastic strain limits that must not be exceeded for the PWR to perform its design function of restraining the whipping pipe.

PWRs are designed to the rules of SRP 3.6.1 and 3.6.2, and to AISC N-690.

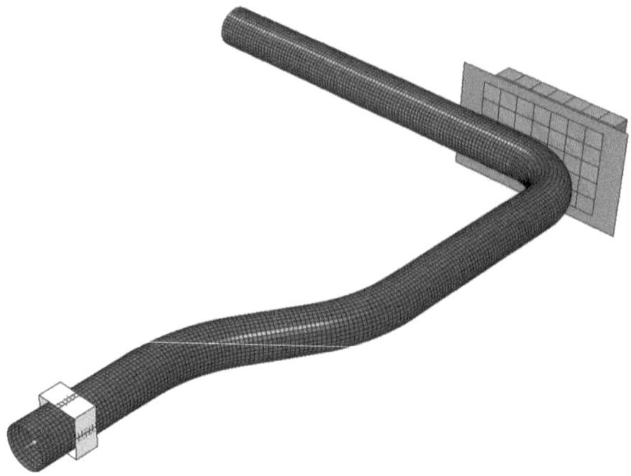

Figure 2.5 Finite element model of pipe whip against a crushable energy absorber.

Figure 2.6 Bumper pipe whip energy absorber.

Figure 2.7 U-bar pipe whip energy absorber.

In some cases the pipe is permitted to whip unrestrained, provided the target, for example, a concrete wall, is able to sustain the impact, with the crushing of the whipping pipe itself absorbing part of the total energy. In the case of a wall, the local effects (spalling, scabbing, penetration, etc.) as well as the global effects (forces and moments imparted to the wall) of the whipping pipe must be investigated.

What are honeycomb whip restraints?

Honeycomb crushable whip restraints consist of sheets of stainless steel brazed together into honeycomb-shaped pillows. The dynamic crush strength of honeycomb is in the order of 5–10 ksi. An issue arose in the late 1970s and early 1980s when it was found that the actual crushing strength of the material was weaker than the values used in the design of whip restraints. This condition prompted the industry to take action, revise the whip restraints design analyses, and make changes as necessary (IN 85-05.)

What is the zone of influence (ZOI)?

The ZOI of a pipe whip and jet impingement is the arc swept by the whipping pipe, with at its broken end the jet. The ZOI is determined by the location of a hinge point along the broken pipe system, which is determined based on the thrust force and the ruptured pipe's material dynamic yield strength. Keep in mind that each break causes two ZOIs: one for the pipe upstream and one for the pipe downstream of the break point.

The ZOI can be as large as 180° from the break point, unless an intervening SSC, such as a substantial structure, a wall, or another pipe of equal or larger size and thickness, is impacted in which case the ZOI for pipe whip would be limited by the intervening SSC. Further hinging (a secondary hinge) may form after impact with the intervening SSC. The interaction should also be evaluated to ensure that the whipping pipe is indeed arrested and the strike is not a glancing blow. The ZOI may also be limited by the fluid inventory feeding the break. For example, if the break occurs near a closed valve, and if the valve remains closed after the break, then there will not be enough fluid between the break and the closed valve to cause the whip. This condition is sometimes known as a limited inventory break.

What is the shape of a jet emanating from a break?

There are basically three types of jets to be considered based on the system thermodynamic condition.

- There is the cold water jet (i.e., nonflashing water) with a ZOI that is essentially an axial extension of the pipe diameter, bending down by gravity. There is no axial limit to this jet zone until it reaches a significant intervening structure such as building concrete or steel.
- There is the ANS 58.2 jet. This jet has three regions, with a radially expanding shape. Historically, this ANS jet has been truncated at 10 pipe diameters following NUREG/CR-2913, by Sandia National Laboratories "Two Phase Jet Loads," 1983. The main steam and MFW lines, for example, would tend to fall under the NUREG jet type.
- There is the more recent (2000s) jet shape that we may call the ACRS jet because it resulted from the questioning of the ANS jet shape by the ACRS. As a result of the ACRS exceptions to the ANS 58.2 conical jet shape, a new spherical equivalent jet was developed and documented in NEI 04-07 to address the blowout of insulation. The spherical shape also accounts for reflections between the incident and nearby SSCs, ignoring shadowing effects.

Note that in addition to the jets, the ACRS pointed out that there is a sonic blast pressure wave that emanates from postulated ruptures in steam lines and will reflect on nearby SSCs of large surface, causing blast forces on these surfaces. This blast wave emanating from the break has been addressed explicitly in some of the newer plant combined licenses.

How were the PRHA methods and criteria developed?

Early in the design of nuclear power plants, the interest in pipe breaks originated with the need to postulate a break so as to hydraulically size the ECCSs and to determine the highest pressure that can be achieved inside containment. The analysis of breaks then evolved to encompass dynamic effects such as pipe whip and jet impingement, environmental effects, and flooding effects. These new regulatory expectations were spelled out in 1974, in Regulatory Guide 1.46, later replaced by SRP Section 3.6.

The nuclear power industry has studied the dynamic effects of postulated HELB for the past 45 years. These studies have included theoretical (analytical) methods, experimental programs, and numerical research (FEA and CFD analysis). It is worth taking a moment for a brief history of the early developments in this field.

- The early pioneering works in the field included a 1965 study on two-phase blowdown from pipes and a 1969 report on blowdown and jet forces, both by GE's Dr Fred Moody and the early experimental work by Faletti and Moulton in 1969. This was followed by more experiments aimed at understanding the dynamic transient phenomena that take place when a break occurs in a high-pressure and high-temperature pipe.

- In 1975 Tractionel (Belgium) conducted tests on the energy-absorbing capabilities of stainless-steel U-bars, compression copper bumpers, and cellular concrete. From 1974 through 1986 the Japan Atomic Energy Research Institute (JAERI) conducted a series of jet discharge tests and studies on 4, 6, and 8 inch pipes under PWR and BWR primary loop pressures and temperatures. The tests were also instrumented to measure target temperature. In 1979, Westinghouse conducted tests on pipes used as elements of PWRs. In 1980 and 1981, Studsvik conducted large-scale jet impingement tests at the Marviken plant in Sweden. The pipe sizes tested ranged from 8 to 20 inch, with pressures up to 700 psi, and fluid ranging from subcooled water to steam.
- In the mid-1980s JAERI also conducted tests of pipe whip with U-bar whip restraints. The whipping pipes were 4 and 6 inch, pressurized at BWR as well as PWR pressures and temperatures. The tests investigated the dynamics of the whipping pipe and the formation of a hinge at the whip restraint. Also in the mid-1980s the Atomic Energy of Canada and the Electric Power Development Company of Japan conducted pipe-on-pipe tests to investigate the effect of 3 inch and larger pipe on 2-inch pipe targets at 1300 psi and 550 °F.
- In 1981–1984, the CEA-CEN (France) conducted tests and studies on the effects of pipe-on-pipe impact using 4-inch whipping pipe on 4- and 2-inch pipe targets. The testing program also included pipe whip impact against steel plates and concrete slabs. The test conditions were those of a PWR at 2400 psi and 600 °F.
- In 1983, Combustion Engineering performed tests on energy-absorbing stainless-steel honeycomb material.
- In 1984–1987, the Pacific Northwest Laboratory conducted for the NRC a series of tests where 6-inch pipes were catapulted against 3–12-inch pipe targets, to investigate the potential for propagation of breaks by pipe-to-pipe impact.
- In 1985 the CEGB and Magnox Electric (UK) conducted pipe whip tests on cantilevered pipe to determine the influence of several factors on the ZOI swept by the whipping pipe. Factors investigated included the direction of the thrust force at the broken end, strain hardening of

the pipe material, whipping pipe with multiple bends, and the difference in behavior between opening and closing bends.
- In 1986, in an EPRI-sponsored study, the experimental results for two-phase jets from the Marviken tests were compared to numerical simulations (EPRI NP-4362).
- In 1988 a revised ANS 58.2 was issued, replacing the 1980 edition, introducing several changes as well as the use of the LBB method.
- In 1990, Siemens conducted tests to investigate what happens at the hinge section of a whipping pipe. The hinge section was also tested with a circumferential crack, to investigate the case of a pipe break occurring in a pipe that had a preexisting crack at the buckle section.
- In the mid-1990s the University of Manchester (UK) conducted experimental and numerical studies of the plastic behavior, ovalization of the cross-section, and buckling of whipping pipes as a function of D/t (the ratio of their diameter to their thickness). The flow restriction caused by ovalization is also addressed in the study.
- The 2000s were marked by the study of jet pressure effects on insulation, the concern being the fragmentation of insulation that would find its way into the sump, and possibly block the recirculation flow (GL 2004-02). This question was investigated by the industry through NEI and Owner's groups, and by the NRC and ACRS, culminating in the spherical jets of NEI 04-07.

How to summarize the key aspects of PRHA?

PRHA is a method of analysis and qualification of postulated pipe ruptures, which follows the criteria of SRP 3.6.1 and 3.6.2.
- Two types of ruptures are postulated: (1) breaks (circumferential and longitudinal breaks) and (2) cracks (through-wall and LBB leakage cracks).
- Three types of effects must be analyzed: (1) dynamic effects, (2) environmental effects, and (3) flooding effects, where flooding can also be viewed as an environmental effect.
- Within dynamic effects, four types of effects must be analyzed: (1) pipe whip, (2) jet impingement, (3) blowdown water hammer, and (4) subcompartment pressurization.

These analyses are complex because they involve dynamic, nonlinear, and plastic effects, they cause many interactions with surrounding SSCs, and they result in large loads on backup structures, possibly the largest magnitude loads on structures.

2.6.4 Leak Before Break

What is leak-before-break (LBB)?

LBB is an engineering analysis process to investigate whether a leak from a cracked pipe can be detected by the operators and isolated, before it propagates into a break.

What are the advantages of LBB?

To answer this question, we have to remember that US NRC 10 CFR 50 GDC 4 requires that "structures, systems, and components shall be appropriately protected against dynamic effects, including the effects of missiles, pipe whipping, and discharging fluids, that may result from equipment failures and from events and conditions outside the nuclear power unit." This requirement is further expanded in SRP Sections 3.6.1 and 3.6.2 which require breaks to be postulated in HE (lines operating at pressures greater than 275 psig or temperatures greater than 200 °F). These postulated, hypothetical, breaks result in significant challenges to the power plant which can be subdivided into environmental effects of the discharging hot fluid, and dynamic effects caused by the sudden break of the pipe. This is discussed in more details in the section of this chapter on PRHA.

GDC 4 goes on to say: "dynamic effects associated with postulated pipe ruptures in nuclear power units may be excluded from the design basis when analyses reviewed and approved by the Commission demonstrate that the probability of fluid system piping rupture is extremely low under conditions consistent with the design basis for the piping." This means that if LBB can be established for a HE line, the dynamic effects of postulated breaks in this line can be excluded from the plant design. The environmental effects, including flooding, and the core cooling consequences from breaks must still be analyzed, without credit for LBB. These advantages and limitations of LBB are addressed in SRP Section 3.6.3.

It seems that regulations (GDC 4 and SRP 3.6) play an important role in defining the rules for LBB?

That is correct, and is due in part to the fact that the LBB method is not addressed in the design codes, ASME III or ASME B31.1. It is primarily governed by regulatory requirements. Some useful references, in addition to GDC 4 and SRP 3.6.3, are:

- NUREG-1061Volume 3, "Evaluation of Potential for Pipe Break," 1984

- NRC Inspection Manual Part 9900 10 CFR "Guidance, Definition of Leak-Before-Break Analysis and its Application to Plant Piping Systems"
- NUREG/CR-4572 "NRC Leak-Before-Break Analysis Method for Circumferentially Through-Wall Cracked Pipes Under Axial Plus Bending Loads," 1986

Is there much of an advantage in eliminating dynamic effects but still having to contend with environmental and core cooling effects?

The dynamic effects from postulated breaks cause significant engineering analysis work and changes in layout, and the addition of braces and whip restraints, adding to the plant congestion and the risk of binding of hot lines by interference from whip restraints. LBB, if it can be established for a given line, would eliminate these efforts and risks.

Can LBB be applied to the reactor coolant system?

The RCL was the first system to which LBB was applied because of three main reasons:

- Nuclear power plants are designed and well equipped to detect a leak from the RCL, even if the leak is quite small.
- The postulated breaks in the RCL have great effects on the reactor pressure vessel, the vessel internals, and the fuel assemblies
- The RCL whip restraints are large, cumbersome structures that hinder access for inspection, and can cause unwanted interference with the expanding RCL during plant start-up.

For all these reasons, LBB was applied to the RCL as a priority, ahead of other lines.

What is the general procedure for LBB analysis?

The LBB effort involves the following general steps:

- Assemble the design and operating (normal operation and transients) data for the candidate line.
- Check whether the candidate line meets a series of prerequisites for applying LBB, which will be described below
- Analyze the line to develop loads, movements, and stresses for all operating modes, including seismic.
- Determine the leak detection capability of the containment.

- Based on the applied loads, postulate a leakage crack size that can be detected by the leak detection capability, with a safety factor of 10.
- Determine if a crack twice the size of the leakage crack is stable under the normal operating loads plus the seismic load, added absolutely.

What are those prerequisites to applying LBB?

Prerequisites are a first screen, to decide whether a line can be considered a candidate for LBB. These prerequisites are stated in SRP 3.6.3. There are over 20 of them. Here are some:

- LBB is limited to HE lines 6 inch nominal pipe diameter and larger
- LBB applies to an entire piping system between anchor points, rigid penetrations, or equipment nozzles; it does not apply to single locations or areas in a line.
- LBB does not apply to cast stainless steel, although additional research and safety margins are being considered as an alternative to this restriction.
- LBB does not apply to lines subject to wall-thinning mechanisms such as corrosion, erosion–corrosion, or erosion–cavitation.
- The LBB lines must be subject to chemistry control in service.
- LBB does not apply to lines subject to SCC.
- LBB lines must be qualified for thermal-induced fatigue.
- LBB analysis must address the reduction of toughness (embrittlement) of the base metal and welds as a result of thermal aging for the plant design life.
- LBB lines must be protected against indirect causes of failure such as fires, missiles, and SSE seismic interactions.
- Water hammer must be prevented in LBB lines.
- LBB lines must be either not prone to stratified flow, or, if prone to stratified flow, the loads, movements, and global and local stresses due to stratification must be included in the LBB analysis.
- The stainless-steel LBB lines must be welded using gas tungsten arc (TIG) welding methods, and the welds must be examined to Class 2 criteria, as a minimum.
- The LBB piping systems must receive 100% preservice inspection, in accordance with ASME III NB-4000 for Class 1 pipe, and NC-4000 for Class 2 and Class 3 pipes.
- The LBB lines must be protected from thermal shock through controlled cooling rate.

- Snubbers in LBB lines must be inspected as part of the ISI and test program.
- The LBB lines must be subjected to augmented ISI.
- The as-built material strength and toughness must be reconciled with the data used in the LBB analysis.
- The LBB lines must be inspected for vibration during the plant preoperational start-up testing.

A unique aspect of LBB is the leak detection capability of the containment. Are there guidelines to determine this leak detection capability?

The LBB analysis is typically based on the capability of a plant's leakage detection system to detect leakage from a small crack. Typical calculated leakage crack size for LBB analysis discharge is 20–40 l/min (5–10 gpm). Since the LBB analysis incorporates a margin of 10 between the calculated leakage and the capability of the leakage detection system credited in the analysis, the leakage detection system has to be capable of timely detection of a leak of 2–4 l/min (0.5–1 gpm).

The leak detection system must meet the performance requirements from RG 1.45. The means of leak detection must be redundant and meet several quality requirements; this explains why LBB is limited to lines inside the containment building. The leak detection systems that are typically credited include:

- The containment sump monitors
- The RCS inventory balance
- The containment radiation monitors
- The containment atmosphere monitors (pressure, temperature, humidity)

Other standards and references related to leak detection capability for LBB applications are:

- NUREG-0800-SRP, Section 5.2.5 "Reactor Coolant Pressure Boundary Leakage Detection"
- NUREG/CR-6861 "Barrier Integrity Research Program"
- IEEE Std 603 "Criteria for Safety Systems for Nuclear Power Generating Stations" Section 5.7 Tests and Calibration
- ISA-S67.03-1982, "Standard for Light-Water Reactor Coolant Pressure Boundary Leak Detection"
- ISA 67.04.01 "Setpoints for Nuclear Safety Related Instrumentation"
- EPRI 1012947 "Materials Reliability Program: Survey of On-Line PWR primary Coolant Leak Detection Technologies" (MRP-187)

Having passed the prerequisites, and having established the leak detection capability, how is the postulated leakage crack determined?

The leakage crack is a postulated crack that, under normal operating loads (pressure, weight, thermal expansion), will discharge a detectable amount of fluid.

The leak rates are calculated based on the flaw geometry, the applied loads, and the crack morphology. The crack opening displacement and crack morphology are typically based on two cases: (1) a fatigue-induced transgranular crack and (2) a corrosion-induced intergranular crack. The leak flow through a crack must be analyzed with the correct thermohydraulic model that captures, as the fluid progresses from the inner diameter (ID) to the outer diameter (OD), a subcooled phase (pressurized hot water) at the ID entrance into the crack, a steam formation phase somewhere between the ID and the OD, and then a two-phase flow region toward the OD, and discharge into the containment environment.

Then, the LBB analysis must postulate that a seismic event occurs while the crack is leaking?

Yes, and crack stability must be established when the critical crack size under maximum normal and seismic loads applied to the cracked section, combined by absolute sum, is at least twice the size of the leakage crack size, to satisfy a margin of two on crack size.

Are fracture mechanics analyses used to determine the critical crack size for LBB analysis?

Yes, and the approach depends on the material under consideration:

- For ferritic steels, the local mode of failure addresses crack tip behavior: blunting, initiation, extension, and instability, using the J-integral method. In this analysis the lower-bound J-R curves must be established using a method approved by the NRC.
- For austenitic stainless steels, the global mode of failure addresses the behavior of the net section, initial yielding, strain hardening, and plastic hinge formation.
- For austenitic stainless steel and their gas tungsten arc welds, the global failure mechanism is evaluated using the limit load method.

From these evaluations, a critical crack size will be determined. A crack larger than the critical crack size would have unstable growth characteristics. A fatigue analysis must establish that the leakage crack, including its future growth, will still have a margin of two against the critical crack size. The minimum tensile properties and the fracture toughness properties used in the LBB analysis must be confirmed through material-specific tests.

Can LBB be applied at locations of weld overlays?

Weld overlay have been used at some locations in the reactor coolant system to provide resistance against SCC. The LBB leakage crack morphology for a weld overlay would be more torturous than for base material, thereby reducing leakage for the same crack size when compared to a crack through the base metal. Therefore leakage detection would be reduced and a greater crack opening and size would be needed for acceptable leak detection with weld overlay.

So, in summary, what are the safety factors in an LBB analysis?

The safety factors in LBB are (1) a safety factor of 10 on calculated leak rate vs leak detection capability, for the leakage crack under the normal operating loads; (2) a safety factor of 2 between the leakage crack and the critical crack size which would be unstable under the absolute sum of operating and seismic loads; (3) safety factors inherent to the fracture analysis; (4) redundancy and quality of the leak detection system; and (5) prerequisites for candidate LBB lines.

CHAPTER 3

Examination and Leak Testing

Chapter 3 addresses the examination of pressure boundary joints, the difference between examination and inspection, and the pressure and leak testing of mechanical systems.

3.1 EXAMINATION

What is the purpose of examinations?

Examination is the name given in the ASME codes (such as ASME III NB/NC/ND-4000 and ASME B31.1 Chapter VI) to the quality control of the shop fabrication and field construction processes, which is done by means of nondestructive examinations (NDE), also referred to as nondestructive testing (NDT). Examinations are different from inspections, but oftentimes codes, standards, and practices interchange the words examination and inspections.

- Regarding examinations, ASME nuclear quality assurance (NQA)-1 has a good definition of quality control examinations (QCs) as "The comparison of the physical, chemical, or other characteristics of a material, component, part, or appurtenance to specified acceptance standards." Another definition is provided in ASME III NCA-9000 Glossary, where examination is defined as "Specific actions by qualified personnel using qualified procedures to verify that items, fabrication processes, and preservice requirements are in conformance with specified requirements." This straightforward definition now gets a second meaning in the rest of the same NCA-9000 paragraph, which continues by saying: "This term [examination], when used in conjunction with qualification of personnel to perform quality-related activities, shall mean a written examination."
- Regarding inspections, in the context of ASME B31 it is the oversight activities by the owner or a representative of the owner. ASME III NCA-5200 expects the inspector of new fabrication or inspection to perform "the detailed inspections in compliance with this Section" and "witness or otherwise verify all examinations and make all inspections required by this Section," and duties are assigned to the Authorized Nuclear Inspector (ANI). In other words, in the construction

ISBN 978-0-12-417248-7

codes, inspection is an oversight function. In ASME XI "Inservice Inspections," inspections are the periodic NDEs performed to check for damage or degradation of pressure-retaining components or their supports. The periodicity of inspections used to be fixed intervals, until the development of risk-informed in-service inspections starting in the 1990s. Article IWA-2000 is entitled Examination and Inspection, and it assigns duties to the Authorized Nuclear Inservice Inspector (ANII).

Is NDT the same as NDE?

Testing in often used to describe methods of examination, such as magnetic particle testing (MT), radiographic testing (RT), etc. Testing also refers to a pressure or leak test, or to the functional test of an active component like a pump or valve operator. To prevent confusion, it is best to describe which technique we are referring to when we call out NDE or NDT.

What are the common examination techniques for systems and components?

There are two large categories of examination techniques: surface examinations and volumetric examinations.

- Surface examinations consist of techniques that probe the condition of the surface of the base materials and the joints. The common methods of surface examination of nuclear components are: visual testing (VT), liquid penetrant testing (PT), and magnetic particle testing (MT). PT is well suited to detect cracklike surface defects and works on all metals. Fluorescent PT used with a dark background is quite sensitive. The caution with PT is that it will give indications even for shallow, inconsequential surface finish discontinuities. MT works on magnetic materials (carbon steel, not stainless steel), and is also good at detecting cracklike surface defects. Another surface examination technique is replication, which is used to make a casting of the microstructure of the metal surface; it is commonly used to detect creep damage in fossil power plant equipment operating at temperatures in the creep range, above the typical 650 °F maximum operating temperature of light water reactors.
- Volumetric examinations consist of techniques that probe the condition of the volume of base materials and welds. The common methods of volumetric examination of nuclear components are: radiographic testing (RT) either by X-rays or gamma rays, including digital radiography; ultrasonic examination (UT), which can be straight beam to measure

thickness or angle beam to search for cracklike defects, also called linear or planar indications; and eddy current testing. Advanced UT techniques include time-of-flight diffraction and phased array ultrasonic testing. Other volumetric techniques include acoustic emission, guided wave techniques, magnetic flux leakage, and electromagnetic acoustic transducer techniques.

In the nuclear power industry, how do today's NDE techniques compare with those implemented 40 years ago?

NDE techniques have greatly improved in the past 40 years, to the point where flaws are detected that would not have been detected in earlier inspections. This is, for example, the case when inspecting steam generator (SG) tubes. In the 1970s SG tube inspections were performed using single-frequency eddy current bobbin probes, which did a good job for wall thinning and longitudinal defects; but the process was elaborate and slow, limiting outage inspections to 5–20% of the tubes at a time. When the eddy current bobbin coil probe is operated in the differential mode it can detect cracks and pits, but may miss gradual wall thinning, such as would be caused by wear. This gradual wall thinning would be detected with the probe operating in the absolute mode. Then, in the 1990s there were several cases of stress corrosion cracking (SCC) of Inconel 600 SG tubes that had to be repaired, and new, more accurate and more efficient techniques were developed. New motorized rotating probes with orthogonal coils are used today to detect cracklike flaws in both directions, circumferential and longitudinal, as well as dents and gouges (IN 13-11).

Why are examinations performed?

There are basically three uses to examinations: quality control, inspection for damage, and forensic investigation.

- Quality control: NDEs are performed as a measure of quality control of new fabrication and construction, to assess the quality and workmanship of base metals and welded joints.
- Inspection for damage: NDE techniques are also used to search for and characterize degradation and damage mechanisms, such as wall thinning caused by erosion or corrosion, or cracking caused by corrosion or fatigue. In this case NDE is either as part of the periodic in-service inspection program performed at each nuclear power plant or as a condition assessment following an abnormal condition such as an overload or an

instance of accelerated corrosion. In this case NDE is part of the inspection program.

- Forensics: Destructive examinations, which involve modifying, cutting, or mechanically testing the material, are used as a forensic tool to determine the cause of failure or the extent of an adverse condition. For example, a component with a fatigue crack may be removed and cut open for visual observation of the crack surface by surface techniques (PT or MT), or metallographic microscopy, with a visual microscope, a scanning electron microscope (SEM), or a transmission electron microscope.

Is it true that class 1 base metal is volumetrically examined?

Base metal defects such as embedded folds, laps, seams, and nonhomogeneities, which cause laminar separation or cracking, will weaken the component. This is why the examination of base metal of Class 1 piping was initially required in 1973, through Regulatory Guide 1.68, "Nondestructive Examination of Tubular Products," introduced as a supplement to NB-2550, in accordance with ASTM E2132, "Standard Method for Ultrasonic Inspection of Metal Pipe and Tubing for Longitudinal Discontinuities," and the additional requirement that ultrasonic examination be performed in the axial and circumferential directions. The Regulatory Guide was withdrawn in 1977 when the volumetric examination of Class 1 materials was adopted in ASME III, requiring the volumetric examination of base metal and weld.

In one case, spool pieces were fabricated on-site for a replacement 18-inch ASME Class 1 seamless pipe, SA-333, Grade 6. When the pipe was volumetrically inspected using SA-655, a lamination was discovered, running around the full circumference. The lamination was oblique, part through-wall. This lamination was missed by the material supplier who only scanned the material in the circumferential direction (IN 84-63).

How about class 1 welds?

The welds of the reactor coolant system boundary (Class 1) are critical and many are difficult welds to make, owing to the bimetallic welds carbon–stainless steel, and the thick walls of the reactor vessel and SGs. For example, welding of the reactor vessel head to the head penetrations (control rod drive mechanism (CRDM) and pipes) is particularly difficult because of the congestion; the need for multiple passes through the thick vessel head; the buttering of carbon steel to stainless-steel welds with nickel alloy (such as Alloy 52M on some of the more recent vessel heads); the need to maintain welding parameters and shielding gas during lengthy multipass welding; the exacting preheat, interpass heat, and postweld heat treatments without deforming the components; and the need for NDE of the welds following the deposition of every few weld layers (IN 13–21).

What standards govern examinations?

We have to differentiate the two parts of examinations: (1) How they are performed and (2) how the results are evaluated and judged. In the nuclear power industry the "how to perform" an NDE of safety-related components is addressed in ASME V "Nondestructive Examination." ASME V specifies how each NDE technique is to be implemented. The qualification of personnel performing NDE is addressed in the construction code itself.

The evaluation of NDE results depends on whether it is an NDE for new construction (QCs) or in-service inspection for the detection of damage.

- For QC of new construction, the results of examinations are evaluated using the construction code. For piping, pumps, valves, and vessels, it would be ASME III NB/NC/ND-4000, for components' supports it would be ASME III NF, for ducts it would be ASME AG-1, etc.
- For service-induced defects in pressure boundary components the evaluation criteria would be those of ASME XI Inservice Inspection.

What are the ASME III division 1 nuclear components requirements for examination of welds?

The examination requirements for welds are spelled out in NB/NC/ND-5000. The fundamental points of examinations are as follows:

- The NDE techniques must follow ASME V. The type of NDE required by ASME III is VT, RT, or in some cases UT, PT, or MT.
- They have to be performed in accordance with written procedures that have been demonstrated to be valid, and technicians performing NDE must be qualified in accordance with standards called out in NB/NC/ND-5000.
- The piece must be cleaned after NDE.
- NDE must be performed after heat treatment with the exception of P-No.1 (carbon steel) where it may be performed before heat treatment. While this exception is permitted for carbon steel, in practice it is much preferable to NDE after heat treatment in all cases.
- Longitudinal welds in pipes, pumps, and valves must undergo RT.

- Class 2 girth (circumferential) butt welded joints in pipes, pumps, and valves must undergo RT. Class 3 girth welds greater than nominal pipe size (NPS) 2 must be RT, PT, or MT. Some instrument tube butt welds can be PT.
- Fillet and partial penetration welds in pipes, pumps, and valves must undergo PT or MT.
- Branch connections and nozzles exceeding NPS 4 must be subjected to RT.
- The outside of welds on branch connections and nozzles NPS 4 and smaller must be subjected to PT or MT, and the inside weld surface as well if it is accessible.
- Welded attachments to pressure boundary must undergo PT or MT.
- The acceptance criteria for examinations depend on the examination technique (RT, UT, PT, MT). In all cases cracks are unacceptable, and in this case the weld must be ground out and repaired. Other indications have acceptance limits.

What types of defects are detected through volumetric examination of welds?

First, let us clarify the use of terms indication, imperfection, and defect:

- Indication is a nonjudgmental term meaning something to report, evidence obtained by NDT. It is what you expect the NDE report to document.
- Imperfection is a discontinuity or irregularity that is detectable by the NDE method used.
- Defect is an imperfection that does not meet the acceptance criterion of the code, the standard, or the engineering design, and is therefore rejected.

The types of imperfections that are typically detected by volumetric examination are illustrated in Figure 3.1. They include cracks (weld, heat-affected zone (HAZ), lamellar along the HAZ, underbead, or base metal), lack of fusion, lack of penetration (incomplete penetration), inclusions (slag, tungsten, or otherwise), porosities, undercut, underfill (concave surface), excessive protrusion on the inside diameter (ID) or outside diameter (OD) (excess reinforcement, convex surface), spatter, and delamination.

What can be detected in surface examinations?

Surface examinations (typically VT, PT, and MT) detect surface cracks, undercuts, and surface porosities, as well as the overall quality of a weld, its

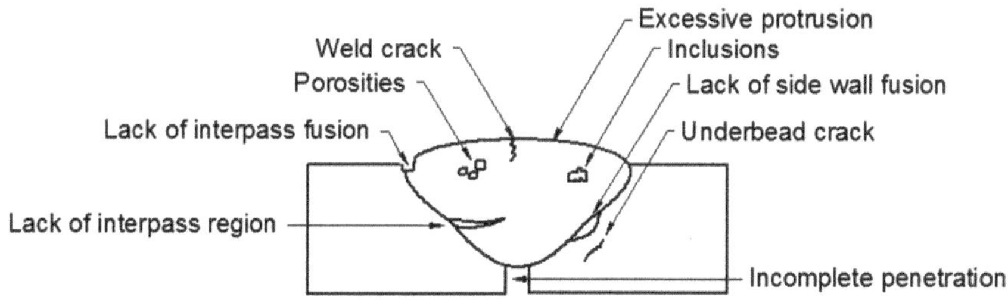

Figure 3.1 Examples of weld defects detected by nondestructive examination (NDE).

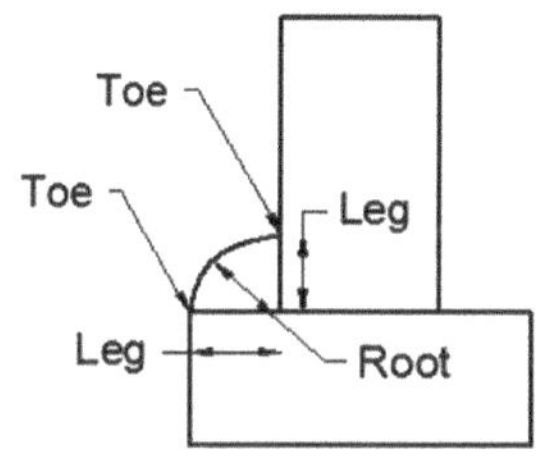

Figure 3.2 Nomenclature for fillet welds.

workmanship. VT is also used to verify the weld size, in particular the leg of fillet welds, Figure 3.2, against the requirements of codes, standards, and the engineering design.

Are examinations also performed as a quality assurance function to verify the material quality?

Checking the quality of materials at procurement is part of the nuclear QA (NQA-1) program. An example occurred when a utility mechanically tested two flanges to find out that the tested mechanical properties, 45 ksi ultimate strength, did not match the manufacturer certified material test report of 77 ksi ultimate strength. In this case the NRC required that each plant identify flanges and fittings from the same supplier and verify the adequacy of their mechanical properties (BL 88-05).

Other forms of examination of the base material are imposed in ASME III NB/NC/ND-2000, and by the material specification ASME II Part D (the ASTM specification). For example, the longitudinal weld seams in pipes must undergo full RT or UT for soundness. There was an instance in the 1970s when a pipe mill discovered that their longitudinal seam welds on SA-312 Type 304 pipe had porosities and lack of fusion defects. The NRC had to request that all plants check the quality of weld seams of safety-related SA-312 Type 304 (BL 79-03).

Which standard governs in-service inspections?

Periodic in-service inspections of safety-related ASME III components are performed in accordance with ASME XI IWA. In-service inspection plans are developed and approved by an ANII. There are strict roles and responsibilities for the owner and inspector. In the United States, the inspector is employed by a state, a municipality, or an insurance company authorized to write boiler and pressure vessel insurance. The inspector can also be employed by an enforcement authority having jurisdiction at the plant. The inspector employer must be accredited by ASME in accordance with the requirements of ASME Qualification for Authorized Inspection-1.

Do plants implement inspections outside of the ASME XI program?

Yes, in particular when generic issues are identified and call for a concerted response. A good example is the augmented inspection of Alloy 600 and their Alloy 82/182 weld metals, which was reinforced by NRC Bulletin 2001-01 (vessel head penetrations), 2002-02 (vessel head and penetrations), 2003-02 (lower head penetrations), and 2004-01 (pressurizer penetrations). Operating experience indicates that this nickel alloy is susceptible to primary water stress corrosion cracking in pressurized water reactors (PWRs). The industry has also addressed this issue through EPRI's Materials Reliability Program which is a nuclear industry program coordinated by EPRI to address materials-related issues. In Bulletin 2004-01 the NRC called for 100% visual inspections of the bare metal pressurizer penetrations for evidence of leakage, and NDE capable of detecting primary water stress corrosion cracking (PWSCC) in case of evidence of leakage. In this case, the damage-specific inspections of the reactor–coolant pressure boundary are consistent with ASME XI:

- NRC regulations at 10 CFR 50.55a refers to ASME XI for the reactor coolant pressure boundary (Class 1).
- ASME XI does not permit leakage of the pressure boundary.
- ASME XI Table IWB-2500-1 specifies examination and references IWB-3522 for acceptance standards.
- IWB-3522.1 in turn requires the detection of leakage, which is done either directly or through evidence of residue.
- When through-wall leakage is detected by visual examination assessment standards are provided in IWB-3142, including supplementary surface or volumetric examinations, corrective measures or repairs, analytical evaluation, or replacement.

What was the approach for the inspection of reactor vessel head penetrations?

The reactor vessel head penetrations of PWRs also contain Alloy 600/82/182 to weld the carbon steel vessel wall to stainless-steel piping and components. After cracks were discovered in several vessel heads, the NRC mandated their examinations, with a tight implementation schedule. The inspection schedule was based in part on the number of effective full power years. Eventually many plants replaced their reactor vessel heads, with Alloy 690 with 52/152 welds in place of Alloy 600 with 82/182 welds.

Would the inspections for flow-accelerated corrosion also fall under the category of damage-specific inspections?

Yes, flow-accelerated Corrosion (FAC) inspections are damage specific. The focus on FAC started on December 9, 1986, when a rupture of a main feedwater line pipe resulted in four fatalities. This accident was covered in IN 86-106, the utility's Licensee Event Reports 86-020-00, 86-020-01, 86-020-02, 87-36, and 88-17, and the NRC IE Inspection Reports 50-280/86-42 and 50-281/86-42. Then came Bulletin 87-01 and Generic Letter 89-08 which required a comprehensive assessment of the potential for FAC, which then evolved into EPRI's FAC program which continues to this day. Bulletin 87-01 required several actions which included:

- In-service inspection for wall thinning.
- Technical basis for the selection of inspection locations, accounting for piping material, layout, chemistry and oxygen content of the water, temperature, flow velocity, etc.
- Inspection results and trending.
- Fitness-for-service assessment and resolution of nonconformances.

Generic letter 89-08 pointed out four cases of erosion- and corrosion-induced pipe rupture. The NRC audited implementation of plant inspections for erosion–corrosion and documented their audit in NUREG-1344. In GL 89-08 the NRC reported "that none of these [audited] licensees has implemented formalized procedures or administrative controls to ensure continued long-term implementation of its erosion/corrosion monitoring program for piping and components within the licensing basis. Therefore, you should provide assurances that a program, consisting of systematic measures to ensure that erosion/corrosion does not lead to degradation of single phase and two phase high-energy carbon steel systems has been implemented." These actions constituted the foundations of the EPRI FAC program.

Where there other instances of damage-specific inspections?

Over the years there have been cases of damage-specific inspections. In the late 1980s there were instances where SA193 Grade B6 Type 410 martensitic stainless-steel studs and bolts joining valve body and bonnets were cracked by stress corrosion cracking (SCC). This was caused by the high hardness of the material, upward of 30 Rockwell C (Rc), combined with the large preload stresses from bolting, and the hot water environment. In this case, plants with these types of valve bolts or studs were asked to disassemble the valves and inspect the bolts for evidence of SCC, and hardness-test the bolts before reuse. To prevent this problem, the steel is normally tempered at 1100 °F, with subsequent check that the hardness is below Rc 26, to verify the effectiveness of the tempering heat treatment.

What are the requirements for examination of safety class ducting?

Examination of safety class ducting is addressed in ASME AG-1, it is a visual inspection for the duct, stiffeners, joints, supports, etc. In addition, welding and weld examination is addressed in AG-1 Section AA-6000. The welds are visually inspected for size, profile (throat and convexity), quality of the fusion joint, porosities, slag inclusions, undercuts, and cracks.

3.2 LEAK TESTING

How does the leak test fit with all the other examinations and tests?

We must first differentiate between the various types of tests conducted on a nuclear power plant system.

- The construction completion test consists of cleaning and flushing the systems, conducting the pressure and leak tests (the subject of this section), calibrating instruments, etc.
- The cold functional tests include the containment leak tightness test, electrical tests, valve stroke, pump operability, etc., all conducted at ambient temperature.
- The hot functional test is achieved by running the reactor coolant pump, which will heat and pressurize the primary system. This is the time for vibration monitoring and thermal expansion monitoring as the plant is heated up in steps and the cooled back down.

- The initial core loading test and verification of in-core instrumentation, core monitoring, radiation monitoring, etc. This is followed by initial criticality achieved by control of boron concentration, and low-power core tests.
- The power ascension test, during which the systems are started up in plateaus up to 100% power, with additional core tests and instruments checks at power; this is also the time for the loss of power test and other transient tests.
- The commercial operation, following regulatory approval at the end of all the tests.
- The periodic leak tests which are performed in accordance with ASME XI, or following repairs or modifications.

What is the purpose of leak testing?

Much like NDE, leak testing has two primary purposes: (1) As a quality control of fabrication and construction joints, for new components, as well as repairs and replacements, and (2) as a periodic check of the integrity of joints after they have been in service, or following an abnormal event.

What are the most common leak testing techniques used in nuclear power plants?

The common leak testing technique are described in ASME V, and they consist of sensitive leak tests and pressure tests.

- The sensitive leak test searches for minute leak paths, and can detect very low leak rates through tight pinholes, cracks, or joints. A sensitive leak test is typically conducted under a low differential pressure of the order of 1 atm differential (15 psid), and it uses sensitive detection techniques such as bubble solution, gas spectrometer, or acoustic or conductivity probe.
- The pressure test is performed either with a pressurized liquid (hydrotest) often water, or with a pressurized gas (pneumatic test) often air. The pressure test challenges the strength of the joint as well as its leak tightness. On one hand it is more demanding than the sensitive leak test because it tries to pry open the joint or defect under large pressure; on the other hand it cannot detect the minute leak paths that a sensitive leak test can detect.

Therefore it is better not to try to label the pressure test (hydrostatic or pneumatic) as better or worse than a sensitive leak test; they each have a different purpose.

How sensitive is a hydrostatic test?

Sensitivity is the flow rate of a leak that can be detected by a hydrotest. A hydrotest can detect leaks in the order of 10^{-2} std.cc/sec. This is a good sensitivity, but not enough to qualify the hydrotest as a sensitive leak test in accordance with ASME Section V. For comparison, a well-conducted bubble solution test can achieve sensitivities in the range of or better than 10^{-5} std.cc/sec. However, keep in mind that the hydrotest challenges the joint at a higher pressure than the sensitive leak test.

What are the leak test requirements for duct?

The test requirements for nuclear plant-safety-related duct are addressed in ASME NGA-1 Sections SA-5300, and TA "Field Testing of Air Treatment Systems." Ductwork testing consists of shop testing of subassemblies, and field acceptance testing of the assembled system before turnover to operation. In the case of field testing, in addition to the leak test, there are nearly 15 other acceptance tests for ductwork, which include several performance tests, a vibration test, and a structural capability test.

Why hydrotest?

The simple answer would be that the hydrotest is required by the ASME construction codes for pressure equipment (ASME III and ASME B31.1). But if the hydrotest is required by the construction code it is because it has several advantages that make it a desired test to verify the integrity of a system:

- First, a hydrotest is the opportunity to check the leak tightness of mechanical joints such as flanged joints, threaded connections, and swaged fittings in instrument tubing. Unlike welds, these mechanical joints do not necessarily have the benefit of assembly by technicians with special qualifications (such as the ASME XI qualifications of welders and welding procedures) or the QC benefits of NDE (such as radiography of welds), Figure 3.3.
- Second, a hydrotest verifies that the whole system can sustain pressures above the design pressure before the line is put into service. It is true that the purpose of hydrotest is to check the leak tightness of joints, but in fact the whole system, including the base metal, are tested. Of course a hydrotest is not a proof of design, and a system that is hydrotested at 1.25 to 1.50 times the design pressure is not qualified through hydrotest, since a code-compliant qualification would require a pressure design margin of 3 or more rather than 1.25 or 1.50. Testing at high

Figure 3.3 Mechanical joints (here bolted and threaded). Do not have the benefit of nondestructive examination (NDE) and rely on hydrotest.

pressures, in the range of three times to five times the design pressure, depending on the code, as a proof of pressure design is called proof testing.

- Third, when a hydrotest is conducted at a hoop stress near yield (and few hydrotests are conducted at such large pressures) the welds are mechanically stress relieved. This little known advantage of hydrotesting at high pressure, near yield, is valuable for pressure equipment that will be in a service prone to fatigue or corrosion cracking.
- Fourth, a hydrotest forces the flushing and cleaning of the line, as water is first introduced into the system, with all drains and vents open. This water flows through open drains, and up vent lines, and washes out rust and residue until the water is clear, before the vents and drains are closed, and the line readied for the pressure test. This can be a double-edged sword: if the component is not fully dried after hydrotest, or if the hydrotest water is kept in the component for too long, it would cause the component to corrode. This is why pneumatic testing is sometimes preferred, especially with stainless steel, to prevent contamination by the water, and stress corrosion cracking from chlorides contained in the hydrotest water. There are many examples in the nonnuclear industry where hydrotest water has caused pitting corrosion, stress corrosion cracking, or microbiological corrosion of stainless steel. Regarding chloride stress corrosion cracking of stainless steel, even potable water can contain up to 250 ppm of chlorides, which is nearly

10 times the 25 ppm that should be permitted when hydrotesting type 304 stainless-steel components. Type 316 stainless steel can accommodate larger amounts of chlorides.

- Fifth, a hydrotest is also an opportunity to verify the strength of the supporting system under the full weight of water-filled components. Of course hydrotest should not be used as a substitute to competent design and construction of pipe supports, but it is a sanity check of the support structures.
- Sixth, when a hydrotest is conducted, it is often necessary to close isolation valves at the end of the section being tested. In this case, the test provides an opportunity to verify that the isolation valves are leak tight at a pressure in excess of the design pressure. Leak tightness of valves should have been verified upfront, at the procurement, receipt, or pre-operational test stages; but the system hydrostatic test provides a welcome confirmation of leak tightness of the isolation valves. Sometimes, plugs with a rubber seal are inserted at the end of piping systems to permit testing the system in subassemblies, Figure 3.4.
- Seventh, if we look back at construction shortcomings unveiled during hydrotest (in nuclear or nonnuclear systems) we realize that a hydrotest can uncover a number of unexpected defects. In one case, with a gas pipeline, the body of a valve ruptured during a hydrotest, the failure was traced to a faulty internal weld repair of the valve body; better to discover this defect with a water-filled pipeline than when the pipeline

Figure 3.4 Plug used to isolate and pressurize the pipe.

is placed in gas service. In another case, a weld developed a pinhole leak during hydrotest, and the leak was caused by a through-wall flaw too narrow to be detected by radiography. In another case, the body of a socket-welded elbow leaked during hydrotest from a minute pinhole defect in the base metal forging.

- Finally, a hydrotest is a final transition point, a hold point, before declaring a construction or a repair complete, and the system ready for turnover to operations.

A hydrotest in accordance with ASME XI Section IWC-5200 "System Pressure Tests" is used as a means of periodically checking the pressure boundary integrity, since none of the NDE techniques commonly used to inspect nuclear power plant system cover 100% of the system.

What documents impose the conduct of a hydrotest?

ASME III NB/NC/ND-6000 "Testing" prescribes the methods and criteria for pressure testing, both hydrostatic and pneumatic, as well as the few cases where systems are exempted from testing, for example, in the case of an open discharge, downstream of the last isolation valve. ASME III NCA requires that the hydrotest (or alternatively the pneumatic test) be witnessed by the ANI, with few exceptions. For ASME B31.1 piping, the requirement for hydrotest is contained in Section 137 "Pressure Tests."

The hydrostatic test is also addressed in Regulatory Guide 1.68 "Initial Test Programs for Water-Cooled Nuclear Power Plants," as one step in the closure of construction activities, and before other operational, functional, expansion, and vibration tests are carried out.

Does ASME XI also require a periodic hydrotest to check the integrity of piping systems?

ASME XI article IWA-5000 addresses the general requirements for periodic pressure or leak testing, or for testing after a repair or replacement. It refers to IWB/IWC/IWD-5000 for testing Class 1, 2, and 3 systems. Under Section XI a system leakage test at a pressure corresponding to normal power operation is sufficient in some cases.

One example of periodic leak testing is the case of the reactor vessel head seal leak-off line. This line detects possible leaks between the inner and outer O-rings positioned inboard and outboard of the vessel head bolt circle. The leak-off line is part of the ISI program, and is leak tested at end of refueling outages, in accordance with ASME XI IWB/C/D-5200, and periodically hydrotested as part of the 10-year ISI interval (IN 14-02).

Has the periodic hydrotest caused problems in the past?

The periodic hydrotest in accordance with IWB/C/D-5000 is a good periodic check of the tightness of the pressure boundary of safety class systems and components. There were a couple of problems in the past, but they have been resolved since through improvements in Section XI. One case that comes to mind is an old IWC-5000 requirement that the hydrotest water temperature had to be at least 100 °F. This was a simplistic way to prevent brittle fracture, but was sometimes difficult to achieve. Since then, ASME XI IWC-5000 was revised to explicitly require a check for the prevention of brittle fracture rather than imposing an arbitrary 100 °F.

When can a pneumatic test be used in place of a hydrotest?

The construction codes (ASME III and B31.1) provide options for conducting a pneumatic test in place of a hydrotest. Typically the pneumatic test option is permitted when the introduction of water would harm the system, including the inability to fully dry the system, or, in some cases, when a steam or gas pipe is not supported to withstand the additional weight of water.

How long should the hydrotest pressure be held?

This depends on the type of hydrotest. If each joint will be visually inspected, the test pressure should be applied for as long as necessary to visually inspect each joint for visible leakage, but not less than 10 min. The ASME codes permit weeping at a mechanical joint during hydrotest, such as a weeping at a valve packing or at a flange, provided the packing is tightened during the test, and confirmed to be leak tight by further observation. Generally, the ASME codes rely on lack of visible leakage at system and component joints rather than pressure hold.

In some cases the hydrotest is a pressure hold test: the system is filled and pressurized, and then the pressured is monitored to remain constant to verify that there is no loss of water or gas due to leakage. In this case of pressure hold test, the time at pressure must be determined on a case-by-case basis, depending on the acceptable leak rate, the volume tested, and the ambient temperature fluctuations.

Can we add temporary supports during the hydrotest of steam or gas pipes?

Yes. In the design of steam and gas piping it is necessary to analyze an additional deadweight case when the piping system is at ambient temperature and filled with water, in order to calculate the hydrotest weight loads on support structures and equipment nozzles, and its effect on the pipe stresses. This hydrotest calculation typically assumes that spring hangers are locked by pinning. It is through this additional hydrotest deadweight analysis that the decision is made to add temporary supports, if necessary and to size these temporary supports.

What is the design analysis allowable stress for hydrotest?

The hydrotest stress limit is governed by the plant final safety analysis report (FSAR). Typically, there is a minimum test pressure set by the applicable Code, and a maximum test pressure to prevent yielding. There is also a hydrotest longitudinal stress limit of S_h or $1.5S_h$ depending on the plant.

What other precautions apply to hydrotest?

Nuclear power plants have formal, step-by-step procedures to conduct hydrotests. Prior to conducting a pressure test, and in addition to the pressure limit and the pipe stress limits, the pressure class of the in-line components (valves, instruments, etc.) and fittings (flanges, threaded couplings, etc.) must be verified to be compatible with the test pressure. In addition, the hydrotest loads for standard pipe supports and spring hangers should be limited to the manufacturer test limit, and in some cases the manufacturer permits twice the catalog-rated load for a hydrotest. Note that a test margin to account for instrument inaccuracies is sometimes applied to the test pressure, typically as a 6% increase in the calculated test pressure, Figure 3.5.

Is it necessary to use a relief device during hydrotest?

A relief valve is not required during a hydrotest. But during a pneumatic test the compressor and the tested system must be protected from accidental overpressure by relief devices, either safety valves or rupture disks.

Any practical words of caution?

When conducting hydrotests, be ready:

- Define the roles and responsibilities for the pressure test, in accordance with ASME III NCA and NB/NC/ND-6000.
- Assemble the staff that will be involved in the hydrotest; have a good, step-by-step test plan; and make sure the staff performing the test has a detailed understanding of the plan.
- The ANI must witness the test if the pipe is larger than NPS 4, otherwise the ANI may refer to the test report.

Figure 3.5 ASME III provides requirements for range and calibration of test pressure gages.

- Prepare the test record, and complete entries that can be completed prior to testing.
- Color code the process and instrumentation diagrams (P&IDs) and isometrics to show the test boundaries. Identify fill points, high point vents, and low point drains.
- Conduct a walk down of the system to be tested, to verify the readiness of the system for testing (access to joints, supports, isolated components, etc.).
- Select the water source and verify water temperature (to prevent brittle fracture and dew point condensation) and water chemistry (to prevent corrosion). The water and metal temperature should be nearly equal.
- Verify the pump or compressor test rig, the relief device (required for a pneumatic test), and the location and calibration of pressure gages. Keep in mind that ASME III NB/NC/ND-6410 has limits on the pressure range of analog and digital pressure gages, and their calibration.
- Inform the ANI of the hydrotest plan, and agree on hold points.
- Check the leak tightness of the isolation valves at the ends of the test section before the test, or use plugs (Figure 3.4), blank inserts in flanges, or blind flanges as terminal points. Valves must be tested separately before installation, in accordance with the valve specification, such as

ASME B16.34. Valve test covers the pressure boundary of the valve as well as the seat tightness of its disk, ball, or butterfly.

- Verify that there is no risk of thermal expansion of the trapped test water, or provide a thermal relief device.
- Calculate the test pressure, in accordance with the Code. For example, the current test pressure for a hydrostatic test in NC-6000 is 1.25 times the design pressure, with a 6% overpressure allowance. Correct for hydrostatic head, and apply a test pressure limit to not exceed yield.
- Prepare supports for the test, pinning springs, and add temporary supports in accordance with the engineering design.
- Isolate instruments or other components that cannot withstand the test pressure.
- Temporarily restrain expansion joints.
- Verify that joints are accessible and visible for the hydrotest, as required by the Code, and select the test duration based on accessibility for joints inspections. Decide what to do if some joints are embedded or inaccessible; in doing so, consult ASME III NB/NC/ND-6240.
- Place strainers upstream of equipment to prevent the introduction of debris that will be carried during the filling of the system or components.
- Vent all high points as water is introduced, to prevent having trapped air in the system.
- Define the pressure hold points, and plan to pressurize in steps; avoid a rapid pressure increase from atmospheric to full test pressure.
- As the line is filled with hydrotest water, keep all the vents and drains open to drain the system from construction debris. Only isolate the system when the flowing discharge water is as clean as the fill water.
- Start to pressurize the system, in steps, as defined in the procedure
- Keep the test hydrotest pressure for a minimum of 10 min, and then the pressure can be reduced to the greater of the design pressure or three-fourths of the test pressure. The inspection time should be sufficient to examine all the joints. The joint inspection should be visual for a hydrotest and using a bubble solution or equivalent for a pneumatic test, unless the test is a pressure hold test if permitted by the Design Specification.
- ASME permits a hold time test for joints that are inaccessible for visual examination during the pressure test. The hold time in NC-6000 is 1 h per inch of wall thickness, but no less than 1 h.
- Have a plan ahead of time to respond to a leaking or burst joint, determine how you will isolate the leak and depressurize the line, determine

if spray or flooding from leakage can damage nearby equipment, and take cautionary measures. Also, account for the energy contained in the system during a pressure test, particularly if it is a pneumatic test; rope-off the test area; and shield personnel. Keep in mind that the harm from a pneumatic test rupture can come either from a blast wave or from ejected shrapnel. Each plant has strict, step-by-step pressure testing procedures to prevent and mitigate bursts during pressure testing.

- Define ahead of time which types of leaks can be fixed during the test, and what would be a leak that causes the test to be stopped and the system depressurized. In doing so, consult ASME III NB/NC/ND-6224.
- Upon completion of the test, drain the system according to the hydrotest procedure, clean the system, and dry and isolate the system.
- Document the test in a comprehensive test report.

What are the requirements for leak testing ductwork?

The requirements for leak testing of safety-related ductwork are spelled out in ASME AG-1. In a nuclear power plant, there are several tests that a duct system must pass:

- A quantified leakage test in which air losses through the duct joints are measured and compared to limits set in accordance with the performance requirements for each individual subsystem. This leakage test is conducted in accordance with ASME AG-1 Sections SA-4500 "Pressure Boundary Leakage" and SA-5300 "Pressure Boundary Leakage Testing."
- A structural capability (strength) pressure test to check for distortion or ruptures. This test is conducted in accordance with ASME AG-1 Section SA-5400 "Structural Capability Tests."
- A weld seam strength test for ducts fabricated with a longitudinal seam, conducted in accordance with ASME AG-1 Section SA-5420 "Longitudinal Seam Qualification Test."
- A functional test to check flow rates, temperatures, humidity, etc., conducted in accordance with ASME AG-1 Section TA-3530 "Functions Tests."

CHAPTER 4

Suspended Systems

Suspended systems, also known as distribution systems, consist of piping systems, tubing systems, duct systems, conduit systems, and cable tray systems. The description of these systems and key features of their design are addressed in this chapter.

4.1 PIPING SYSTEMS

Why is the engineering of nuclear plant piping systems a discipline all of its own?

When entering a nuclear power plant, one first notices the large number of piping and tubing systems in all three buildings: the reactor building, the auxiliary building, and the turbine building. Piping systems are the plant arteries; they transfer heat energy from the reactor core to the turbine, directly as in boiling water reactors or through the steam generators as in pressurized water reactors (PWRs). Other piping systems provide normal heat removal during shutdowns, chemical control, blowdown, emergency cooling, and equipment cooling by carrying water to and from the ultimate heat sink (a lake, a river, or the ocean). Then, there are the instrument tubing systems, which are also piping systems. All in all, a 1000-MW PWR has nearly 20 miles (32 km) of pipe, and thousands of valves and instruments.

The design analysis, qualification, and operability evaluation of all these piping systems combine (1) the complexities of structural analysis, (2) the need for pressure boundary integrity at high operating pressures and temperatures and under transient conditions, and (3) the need to maintain operability (flow control) through passive components (the pipes themselves, the support structures) and active components (pumps, compressors, valves, instruments, snubbers, etc.). Some engineers dedicate their career to this field.

Also, if pipes are not correctly designed, fabricated in the shop, constructed in the field, and inspected regularly once the plant is in service, they can fail by one of many damage mechanisms (corrosion, vibration, stress corrosion cracking, overpressure, thermal fatigue, etc.), which jeopardizes operation.

Nuclear Power Plant Safety and Mechanical Integrity
ISBN 978-0-12-417248-7

Is instrument tubing also considered piping?

From a design and construction point of view, they are both covered under ASME III or ASME B31.1. The two can be differentiated by their size, with piping procured by pipe schedule, to ASME B36.10 and B36.19 standards, while tubing is procured custom size, specifying the tubing outer diameter and thickness. Piping and tubing can also be differentiated by their function: conveying fluid in pipes and performing instrumentation and control functions for tubing. By piping system, we mean not only the pipe segments themselves, but also the fittings (tees, elbows, reducers, etc.), the in-line components (flanges, valves, instruments, filters, etc.), and the pipe supports and braces, either standard supports or custom-designed. While piping components are in-line items, the term equipment can be used for self-supported items such as pumps, compressors, tanks, and vessels.

How are the functions of piping systems differentiated?

As we have seen in earlier chapters, all structures, systems, and components (SSCs) in a nuclear power plant are classified according to their safety class and their seismic category. Piping systems, together with their components, instruments, and controls, are considered safety related when they perform one of several key functions: (1) they are relied upon for safe shutdown of the plant, (2) they are relied upon for the removal of the decay and sensible heat from the reactor, or (3) their failure or malfunction could cause a release of radioactivity.

Which design codes apply to piping systems?

The applicable design code depends on the vintage of the nuclear power plant construction permit (CP).

- Pre-1969 CP: Plants with a CP issued earlier than 1969 were designed to ASME B31.1 with supplementary requirements for safety-class systems that were specified in the preliminary safety analysis report and final safety analysis report (FSAR).
- 1969–1971 CP: In 1969, ASME B31.7 was implemented specifically for nuclear power plant piping, replacing B31.1 for safety-related piping systems.
- Post-1971 CP: In 1971, the design and fabrication requirements for nuclear plant piping were folded into ASME III, which until then only covered vessels. Plants with a CP dated after 1971, and to this day, are designed and constructed in accordance with ASME III. Nonsafety piping was and continues to be designed and fabricated to ASME B31.1.

The situation is somewhat similar for pipe supports, which were designed and fabricated to the AISC "Manual of Steel Construction," until pipe supports were added to ASME III, as Subsection NF.

Where is the boundary of a piping system?

The boundary between the ASME III NF pipe support and the building structure, where the support steel ends and the building steel structure starts, is specified in a series of figures contained in ASME III NF. This boundary normally ends at the end of the pipe support attachment to the building steel or concrete. For example, the boundary can be defined to be at the weld of a structural attachment to a baseplate, where the weld would be the last element to be qualified under the ASME III NF.

The boundary between the piping and a vessel, pump, or valve nozzle is (1) the first circumferential weld to the equipment or component, where the weld is considered part of the piping, (2) the face of the first pipe-side flange in a bolted connection where the bolts are considered part of the piping, or (3) the first threaded joint in a screwed nozzle connection.

Are the design codes (ASME and AISC) sufficient to analyze and qualify the large number of piping systems in a nuclear power plant?

No. While the design codes, whether ASME III, ASME B31.1, or AISC, address the qualification criteria and stress limits that must be met, they are not meant to cover all the aspects of a correct and complete design. Instead, the rules for modeling and analyzing piping systems are typically captured in company-specific piping and support design procedures, standards, and guides. These company standards and guides are necessary because the analysis of piping systems involves, as we will see, many facets. A new development took place in the 2007 edition of the Standard Review Plan (SRP), when a new Section 3.12 was added that lists several considerations for the analysis of piping systems.

Where is the starting point for pipe stress analysis?

The starting point in design stress analysis is to decide how the thousands of feet of piping and tubing runs in a nuclear power will be subdivided into manageable subsystems that can be modeled and analyzed, and filed in a retrievable form. This is one of the first steps: subdividing and grouping piping systems for design analysis. One good method for tracking piping analysis packages is to generate one set of plant P&IDs color-marked with the boundaries of the individual stress calculation package numbers.

Clear end points for modeling piping systems are the anchors—points that restrain the six degrees of freedom of motion of the pipe. Other ending points are floor-anchored equipment such as pumps, compressors, tanks, or vessels. Where the end point of a model is a piece of equipment, it is important to address several key aspects of the model:

- Will the pipe model end at the equipment nozzle or will it proceed all the way down to the equipment axis or even its anchorage to the building structure?
- How to account for the global flexibility of the equipment (for example, its bending or rocking flexibility) and the local flexibility of the equipment nozzle and shell?
- How to account for the movements at the equipment nozzle caused, for example, by thermal expansion, seismic motions, or other dynamic motions?
- How to account for the amplification and the frequency shift of the seismic floor excitation as it is transferred from the equipment base, through the equipment itself, to the pipe nozzle?
- Who has the responsibility for qualifying the equipment-to-pipe connection?

Is there another way to subdivide piping systems, other than anchors or equipment nozzles?

Small branch lines can be decoupled from their large headers and analyzed in separate models. The rules for decoupling headers from branch lines are not addressed in ASME B31.1 or ASME III, and they have taken different forms over the years, in different projects. For example, decoupling of large headers and small branch pipes has been based on the ratio of header-to-branch moment of inertia, the ratio of section moduli, or simply the ratio of pipe sizes. A simple and most recent criterion is to permit decoupling when the ratio of header-to-branch pipe sizes is 4:1. Decoupling criteria are also addressed in ASCE-4 and in SRP Section 3.7.2. The above questions for decoupling at equipment nozzles also apply to the decoupling of header and branch lines into separate analyzes. There are additional considerations for the decoupling of branch pipe beyond those listed above for equipment. For example:

- In the analysis of the decoupled lines, it is necessary to include the stress intensification factors or stress indices at the decoupled point, typically a tee or a branch.

- If the branch line has an anchor near the decoupling point, it cannot be decoupled, since the header will be subjected to the restraining action of the branch anchor.
- If the branch line is near an equipment nozzle or a penetration, its effects on the nozzle or penetration cannot be ignored through the decoupling process.
- The weight and inertia effects of the branch segment near the header should be considered in the run pipe model, for example, by adding the mass of half the span of the branch pipe to the centerline of the run pipe.

Is it possible to subdivide a piping model by overlap?

Sometimes, the so-called overlap method can be used to terminate a model into an adjacent model, without the need for a physical anchor. A model that starts at point A and ends at point B can be extended to overlap into the adjacent model up to a point C. Then, the portion B–C serves to reflect the effect of the adjacent model onto A–B. The length of section B–C tends to be large as a minimum of two supports in each direction should be included in the overlap. The overlap method is not a simple or particularly elegant procedure, and wherever possible is avoided in favor of actual anchors.

This subdivision into stress analysis subsystems can result in a large number of analyzes?

Yes, a nuclear power plant has hundreds of piping analysis packages, and it is essential to clearly number, file, and track the analyzes packages to be able to retrieve them for modifications or operability assessments during the plant life. The tracking system should be such that if we were to point to a pipe segment on the P&ID, it would be possible to know which pipe stress analysis package applies to that segment. And, once the piping stress analysis package is retrieved, it should be possible to readily know which pipe support calculation packages apply to that piping model.

What type of software analysis should be used?

An important early decision is to select the analytical software. The software for piping analysis falls into one of two general categories: special-purpose code analysis software and general-purpose finite element analysis (FEA) software.

- The special-purpose code analysis software contains a menu of piping, fittings, and pipe support elements that can be stringed together to build a model. The model is then run, and postprocessing routines are already built in to calculate code stresses and compare them with code-allowable limits. The piping elements have preset code-specified flexibility factors to represent the deformation of elbows and branch connections to account for their cross-sectional deformation. The flexibility factors for fittings are specified in ASME B31.1 and ASME Section III subarticles NB/NC/ND-3600. These special-purpose software are based on elastic analysis but are not necessarily linear. In other words, the stress–strain curve for the material is linear ($\sigma = E\varepsilon$), but the model may have the capability to model gaps or even bilinear support stiffness. Modern code analysis packages have the capability for static and dynamic analysis, either modal analysis (with a linear model) or time-history analysis.

- The second type of analysis uses the general-purpose FEA method. The FEA software can model practically any type of shape or structure, and can perform plastic and nonlinear analysis. The classic cautions for FEA apply here, but, in addition, two questions must be addressed before selecting the model elements: (1) Will the analysis results be evaluated using NB/NC/ND-3600 stress equations (basically beam formulas), or will they be evaluated using the stress linearization of NB/NC/ND-3200 or the nonlinear methods of ASME III Appendix F? (2) Are the fittings flexibilities of NB/NC/ND-3600 apply or will the flexibilities of the model take precedence? For example, will the flexibility of an elbow modeled with finite elements, such as shell elements, be the same as the design code flexibility factors?

Are there established protocols for verifying and validating the piping analysis software?

The formalities for verification and validation (V&V) of safety-related software are addressed in 10 CFR 50 Appendix B. Regarding special-purpose piping stress analysis software, the Nuclear Regulatory Commission (NRC) has published benchmark problems in NUREG/CR-1677.

How important is V&V?

Very important, but it has to be done correctly. The emphasis has to be on the technical intricacies of the software and how it is used. Those verifying and validating software (so-called V&V), and those auditing the V&V of software (so-called QA engineers) have to be experts in the use of the software and its technical basis. Errors in software have occurred in the

1970s and early 1980s. Today, because of decades of experience and corrections, these pipe stress analysis software errors are rare. In all cases, if an error is discovered, all analyzes that used the erroneous software feature are reviewed for the impact of the error on the completed work. One case that comes to mind goes back to the early 1980s; in that case, a piping analysis software calculated stresses in a branch pipe using the larger header pipe section modulus rather than the smaller branch line, and therefore the software underestimated the branch-side stresses (IN 83-31).

In most cases, as all good analysts know, errors are not due to the software itself but rather due to the incorrect understanding and use of the software. This can happen at several levels: First, errors can be made when the analyst does not fully grasp the physics of the phenomena being modeled. Then, errors can occur in modeling, in checking the model sensitivity to modeling assumptions, in applying the loads to the model, in interpreting the results, and in proof-checking the results using sanity checks. This is particularly true for dynamic phenomena such as fast hydraulic transients in accident analyzes. One case in point is the mass and energy release analyzes conducted to calculate the large post-steam line break ambient temperatures and pressures in compartments. In one case, the analyst did not account for the additional superheating of escaping steam from the uncovered, overheated portion of the steam generator tube bundle (IN 84-90).

Which of the two types of software, specialized code analysis software or general-purpose FEA software, is used in practice?

In practice, the special-purpose pipe stress analysis software, rather than FEA, is used for pipe stress analysis because of their simplicity and the embedded code-compliance elastic stress checks. FEA is mostly reserved for the analysis of abnormal events and conditions, and to analyze extreme loads, such as tornado missiles, or large dynamic transients where the plastic option may be required to better understand the system response.

Are all piping systems analyzed?

Most piping systems are modeled and analyzed, but some are qualified using standard span tables and configuration details, often referred to as cookbooks. These standard spans and configurations are prequalified pipe and support arrangements (straight runs, bends, loops, etc.), that when stringed together would result in a prequalified layout. This approach has commonly been applied to instrument tubing and to some small bore piping (nominal pipe size (NPS) 2 and smaller NPS).

What are the key attributes to watch when starting to build a piping stress analysis model?

There are several key attributes to building a good model for piping system analysis and qualification, which explains the need for project-specific standards and guides. These project standards and guides must be agreed upon up front to prevent rework, and cost and schedule overruns. The following list can serve as a road map of the topics to be addressed in project-specific piping analysis guidance:

- A complete set of isometrics from end point to end point of the model, and the boundaries shown on P&ID, including the interfacing adjacent models.
- The safety class, applicable code, and seismic category of the system.
- A complete set of pipe support information (drawings are preferred if available): type, location, direction, stiffness, tributary mass, capacity, gaps, etc.
- The pipe specifications with material, size, types of fittings and components, with their size, pressure class, and weight.
- The insulation type and weight.
- A project-wide numbering sequence for the node points in models has proven to be very useful in the few cases where it was used.
- A project-wide set of coordinates (for example, north is X, vertical is Y, then east is Z, or the like).
- A line list with pressures and temperatures for each operating mode and postulated service level (A, B, C, or D).
- For Class 1 lines, the full pressure and temperature transient in the form of curves of pressure and temperatures as a function of time P(t) and T(t), with the corresponding number of cycles.
- The applicable loads and load combinations, including equipment nozzle or header movements, seismic input, hydraulic loads, if any.

The best piping design, analysis, qualification projects (most cost-efficient and technically sound) have been those with correct, clear and well-illustrated, step-by-step modeling guidelines so that at the end of the day all the stress packages look alike.

How accurate must the piping models be, down to the inch?

A good starting point to address the accuracy of piping system models is ASME III Appendix T, which was developed with as-built reconciliation in mind. As-built reconciliation is the reconciliation of the analyzed and qualified configuration against the one that was installed in the field. This issue became front and center in 1979, with the as-built issue captured in NRC IE Bulletin 79-14.

What was the as-built issue?

As-built reconciliation is the need to have consistency between the analyzed and qualified numerical model of a piping and support system and the installed (as-built) system. This was one of the most important engineering and construction issues faced in the late 1970s in terms of the amount of review and reanalyses the plants had to perform. As each plant embarked on as-built reanalyses, several issues not related to as-built were addressed at the same time, for example:

- The use of erroneous seismic response spectra (IN 79-31).
- Checking reaction loads from pipe supports onto the backup structure that had not been fully analyzed (IN 79-28).
- A memorable case took place in 1979 when the NRC ordered several plants to shutdown within 48 h because of an error in piping analysis software, which surfaced during the as-built reanalyses (IN 79-06, IN 79-36).

Is as-built something that was recognized and resolved in the 1970s?

Well after the issue of IE Bulletin 79-14, "Seismic Analyses for As-Built Safety-Related Piping Systems," the industry periodically reminds itself, and the NRC reminds the industry, of the importance of having correct and complete as-built information reflected in design documents. A case was reported of temporary pipes that were left in place after a refueling outage, rather than replaced as planned (IN 85-66).

What are the methods of seismic analysis and design of piping and tubing systems?

First is the decision of what type of analysis should be performed. Practically, there are three options: (1) Modal response spectra seismic analysis, which is the most common seismic analysis method for safety-class nuclear piping systems, (2) static seismic analysis, and (3) standard span methods (cookbook), which is a common method for small bore piping (NPS 2 and smaller) and tubing systems.

What are the important considerations in modal analysis?

The modal response spectra method relies on the dynamic accuracy of the system mode shapes and natural frequencies. The important considerations are (1) accuracy of the model and (2) the seismic input.

We have addressed model accuracy earlier, but we must repeat a few key points: the importance of the spacing of node points, the stiffness of supports and interfacing equipment, and the modeling of offset weights such as the eccentric weight of a heavy valve operator. The level of dynamic accuracy, for example, the spacing of the model node points, depends on the frequency content of the input. If the in-structure response spectra (ISRS) peak at high frequency, in the range of 30–50 Hz as is currently the case for central and eastern US spectra, then the model must correctly capture these high frequencies.

For each of the three directions (north–south, east–west, and vertical), the seismic input is typically the ISRS that envelope the attachments of all the system supports and anchors. An alternative would be to use what is labeled the multi-input response spectra, where supports are grouped, for example, by elevation, and a different seismic input is applied to each grouping of supports. The multi-input spectra methods and cautions for its use are addressed in NUREG-1061 Volume 4.

The ISRS must be selected at the correct damping, which has changed over the years from 0.5% damping for the earlier plants up to 5% damping today, and in the 1980s several plants applied a frequency-dependent damping published as ASME III Code Case N-411 and has become known as N-411 damping. A word of caution: The damping values in standards such as ASME III Appendix N may be different from those published in the NRC RG 1.61. In all cases, the damping used at a specific plant should be as stated in the plant FSAR.

Once the ISRS are correctly selected, several other modeling and analysis decisions must follow the FSAR commitments, which include the following: the method of modal combination (how to combine modal results, including the results from modes that are closely spaced), the method of directional combination (how to combine NS-EW-vertical responses), the cutoff frequency, the zero-period acceleration and dynamic mass participation, and the rigid-range correction must also follow the methods stated in the plant FSAR. These topics are addressed in NRC RG 1.92.

In light of these difficulties, why not use a static method for seismic analysis?

The static seismic analysis method is common in the seismic design of piping systems in all other industries (fossil power, chemical, above-ground pipelines) that follow the ASCE-7 static seismic design method. It consists of applying a constant acceleration to the distributed weight of the system in each of the three directions: north–south, east–west, and vertical.

In the nuclear power industry, the use of a static analysis method should address the three conditions for static analysis stipulated in NRC SRP Section 3.7.2 and quoted here:

- "Justification is provided that the system can be realistically represented by a simple model and the method produces conservative results in terms of responses. Typical examples or published results for similar structures may be submitted in support of the use of the simplified method."
- "The simplified static analysis method accounts for the relative motion between all points of support."
- "To obtain an equivalent static load for an SSC that can be represented by a simple model, a factor of 1.5 is applied to the peak spectral acceleration of the applicable ground or floor response spectrum. A factor less than 1.5 may be used, if adequate justification is provided."

For the static analysis, the peak spectral acceleration, at the appropriate damping for the operational basis earthquake (OBE) and the safe shutdown earthquake (SSE) is selected. This peak must envelope the ISRS at all attachment points of pipe supports to the building structure, and this acceleration value, in each direction (north–south, east–west, up–down), will be referred to as the "g" input. The g in each of the three orthogonal directions is multiplied by 1.5, which is meant to account for the multiple modes inherent in the seismic response of a piping system. This acceleration is then applied to the distributed weight of the system, including the concentrated weight due to in-line components such as valves.

The third method mentioned is the standard span method. What is it?

Because there are such a large number of piping systems to analyze at the design stage, since the 1960s engineers searched for means to develop and use standard layout and support arrangements that are seismically prequalified. This approach proved to be useful for small bore piping and tubing (NPS 2 and smaller) that did not require a thermal expansion analysis because they are flexible and operate below 150 °F (65 °C) (Figure 4.1). Most nuclear power plant projects developed standard span tables and standard support loads for small bore piping and tubing systems to be applied without the need for a detailed analysis of each system. An example of this approach can be found in EPRI NP-6628 "Procedure for Seismic Evaluation and Design of Small Bore Piping (NCIG-14)." These guidelines containing standard span tables and figures referred to as cookbooks. If the

Figure 4.1 Small bore piping can be supported using standard span tables.

pipe operates at a temperature above 150 °F, it is typically modeled for flexibility analysis, in which case the seismic analysis is also performed on the model, either static (when the above three conditions for static analysis are met) or dynamic.

Is the cookbook method more efficient than analysis?

There is a debate regarding this question. Some designers feel that today's piping analysis software is so good that it is often faster to model and run the piping system than to figure out the spans using a cookbook. Figure 4.2 is an example of a layout that was initially analyzed using a pipe stress model. Then, the routing had to be modified on short notice, and a support had to be redesigned. If the original model cannot be retrieved and verified on time, then hand calculation methods based on spans can be used because the layout is relatively simple. While the pipe layout was simple, the new support structure was sufficiently complex to require FEA of the support itself using an AISC-based software (Figure 4.3). In this case, the software provides the support member and weld stresses, and automatically compares them with code allowables. In addition, and separately, the pipe support designer checks the demand vs capacity of the baseplates, the anchor bolts, and the standard support components (*U*-bolts, struts, snubbers, springs, etc.).

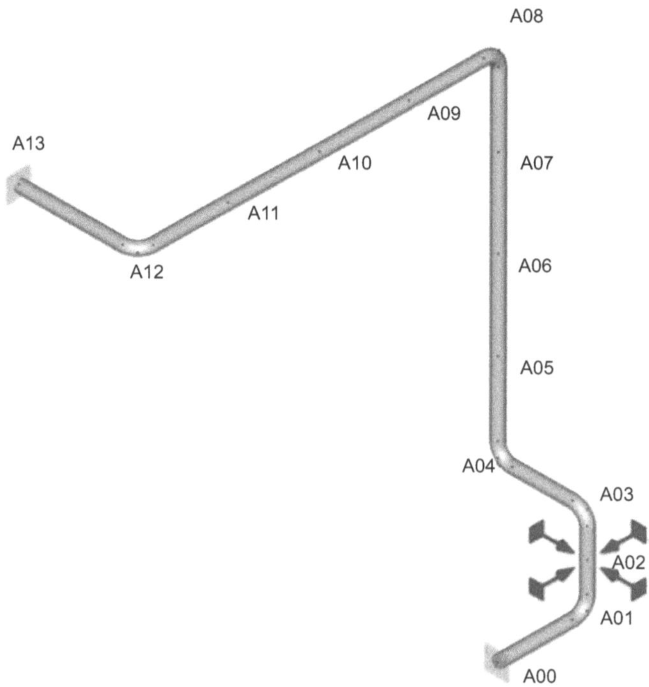

Figure 4.2 Simple layout that can be modeled and analyzed or qualified by cookbook.

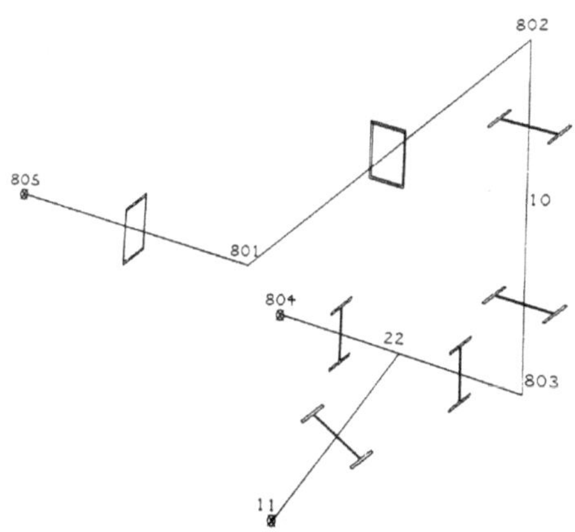

Figure 4.3 Pipe support frame modeled and analyzed for AISC or ASME III NF compliance.

What are the loads and load combinations for the analysis and qualification of piping systems?

In a nuclear power plant, the loads to be considered in design of safety-class SSCs are listed in the ASME III design specifications. The design specifications also specify how the loads should be combined. These load combinations must be consistent with Table I of Appendix A of SRP Section 3.9.3, and they are captured in the plant FSAR under the same Section 3.9.3.

The load combinations are grouped into Service Levels A, B, C, and D. These Service Levels used to be called Normal, Upset, Emergency, and Faulted, respectively. These are defined in ASME III, paragraph NB-3113. The following five plant conditions defined in ASME III are considered in the design of mechanical systems and system supports.

- Service Level A, which is used to be called a Normal operating condition, covers normal system start-up, operation in the normal power range, hot standby, and normal system shutdown.
- Service Level B, which is used to be called an Upset condition, covers a deviation from normal conditions anticipated to occur often enough that the design should include a capability to withstand the conditions without operational impairment. In practice, the upset conditions include the following: transients that result from an operator error; the malfunction of a component requiring its isolation from the system; transients caused by loss of load or loss of power; incidents not resulting in a forced outage; incidents resulting in a forced outage, but for which the corrective action does not include a repair of mechanical damage; a reactor trip from full power; the reduced or partial loss of flow of a critical pump; and the occurrence of an OBE seismic event (five OBE's for a total of 50 cycles of shaking, consistent with SRP 3.9.2).
- Service Level C, which is used to be called an Emergency condition, is a condition that requires shutdown for correction of the conditions or repair of damage in the systems. The break of a small steam line and the loss of pump flow are examples of emergency conditions.
- Service Level D, which is used to be called a Faulted condition, is a condition whose consequences would impair the integrity and operability of the system. Nuclear power plants are designed to safely withstand one of these Service Level D occurrences. Examples of Faulted conditions are as follows: a large line steam break, a large reactor coolant loop break, a single SSE (10 cycles of maximum seismic amplitude, consistent with SRP 3.9.2), and a steam generator tube rupture.

In addition to the four service levels, there is a design load case, typically identical to Service Level A, and a test load case (such as the pressure load case during a hydrotest).

What are examples of loads to be addressed in design?

Loads to be addressed in design consist of weight, pressure, thermal expansion–contraction, anticipated fluid transients, seismic (OBE and SSE), and postulated pipe breaks.

Thermal loads on a piping system include the effect of the pipe metal temperature, the expansion of the equipment to which the pipe is attached, and the building expansion that would result from the postulated breaks inside a compartment. This building expansion causes penetrations and support attachments to move, and as a consequence causes loads on the piping system. In nuclear plants, temperatures greater than 150 °F are considered to be significant for thermal expansion analyzes of the piping system. Piping expansion before reaching this temperature is normally accommodated by the closing of support gaps, contact friction, as well as the flexibility of bends typically encountered in piping systems. Of course if a pipe is laid out straight between rigid anchors, any heating, even below 150 °F, will cause problems; but this is mostly an academic concern, since in practice, piping systems have bends, which give the system its flexibility.

What are the effects of the high operating temperatures on a piping system?

First, there is the expansion and contraction of the piping, what we may describe as the beam effect: The pipe, a series of beams, expands and contracts as the temperature changes from ambient to the hot condition, as high as 650 °F in certain systems, and back to cold at shutdown. All codes, ASME B31.1 and ASME III, address these beam expansion effects.

Operation at high temperature also causes local transient temperature gradients at specific points in the system. These gradient effects are explicitly addressed in ASME III for Class 1 piping, but not for Class 2 or 3, or for B31.1 systems. The designer cannot ignore these effects and argue that "they are not addressed in ASME III NC/ND-3600." For example, there have been cases where a hot water stream mixed with a colder water stream at a tee, in Class 2 and 3 piping. This continuous fluctuation of temperature caused cracks at the tee. The designer should study the P&ID and the system operating modes to assess whether local thermal shocks, thermal mixing, or thermal stratification should be addressed, even if not called out explicitly in the ASME design code.

What are the qualification requirements for piping systems?

Once the computer runs are completed and the loads are obtained and combined, the designer should carefully study the results against the ASME III design-specification requirements for qualification. Practically speaking, here are the parameters that must be qualified:

- The pipe stresses against code limits and project-specific limits
- The pipe movements against interferences
- The equipment nozzle loads against manufacturer limits
- The penetration loads and movements against civil-structural-specified limits
- The loads and movements at supports to be qualified
- The loads and movements at mechanical joints, such as flanges, expansion joints, swage fittings
- The anchor loads, from both sides of the anchor, to be qualified
- The pipe-to-support welded attachments, to be qualified as described, for example, in ASME III Appendix Y
- The accelerations at active in-line components, such as valve operators, against qualified limits
- The pipe break stress limits for high-energy lines against the limits of NRC SRP Sections 3.6.1 and 3.6.2
- The leak-before-break stress limits for LBB lines against the stress limits established in the LBB analysis

What are the allowable stresses service level load combination?

The allowable stresses in each case depend on the type of component and support structure, and the vintage of the design. Depending on the plant vintage, some plants have allowable stresses for piping systems of S_h for design: $1.2S_h$ for Service Levels A (Normal) and B (Upset), $1.8S_h$ for Service Level C (Emergency), and $2.4S_h$ for Service Level D (Faulted). Piping that must perform an active function (deliver, control flow) after a postulated Service Level D accident is in some cases limited to Service Level C stresses, which is referred to as functional capability limit. The stress limits for thermal expansion–contraction stresses are typically the ASME III and B31.1 allowable S_a and have not changed over the years.

There are also stress limits to prevent the postulation of high-energy pipe breaks at high stress points. These limits are typically 80% of the allowable for Service Level B (including OBE) plus thermal stress, and 80% of the allowable for the thermal expansion stress by itself. These stress limits are based on the SRP and FSAR Section 3.6. For the newer plants where OBE is eliminated, the pipe break location stresses are based on the pressure, weight, and thermal expansion stresses, excluding the OBE stresses.

Was there not an issue with the contact stresses between a pipe and its support?

In the nuclear power industry, every question that may affect the reliability of equipment is investigated thoroughly. Most of the time, these are real issues that are resolved through analysis, testing, or hardware modifications. But in a few cases, this relentless pursuit of "what if" has led to tackling academic issues till it became fully and quantitatively proven that they are inconsequential in practice. An example of academic issue was the hunt in the 1980s for "stiff clamps" that, it was thought, could fail a pipe by excessive bearing pressure between the pipe and the clamp (IN 83-80).

So far we have focused on the pipe itself, what about the supports?

It may be good to start with the grouping of support categories. There are (1) the standard catalog supports, (2) the custom-engineered steel frames, and (3) the baseplates, either embedded or bolted to the concrete with cast-in place or expansion anchors.

What are examples of standard supports?

Standard supports are catalog items that are procured with a load rating. They consist of springs, struts, snubbers, vibration dampers, and the standard hardware for their attachment to the pipe (clamps, brackets, clevis, bolts, etc.). They are typically procured to MSS-SP standards, but have to meet ASME III NF for safety-related ASME III piping systems.

- Spring supports normally restrain only the pipe deadweight without restricting pipe thermal expansion (Figure 4.4). Spring supports, also referred to as spring cans, can be (1) variable (most commonly used) where the load exerted by the spring on the pipe changes with displacement or (2) constant-load hangers where the upward load on the pipe remains constant within a range of vertical pipe movements. Constant-load hangers are normally utilized in hot-temperature applications,

Figure 4.4 Variable spring hangers.

where the vertical growth is larger than what a variable spring can accommodate. Note that there have been cases where the constant upward load applied by constant supports onto the pipe drifted over the years. The constant-load hangers need to be tested, adjusted, and eventually replaced periodically. We have seen this in old fossil power plants, and it is also true for nuclear power plants. One plant tested four safety-related constant-load supports on a steam line and found the supporting force had drifted by nearly 10%, more than the 6% permitted in ASME III NF. This was attributed to wear of the mechanical links in the constant-load supports (IN 09-04).

- Struts are rigid bars that restrict the loads acting along their axis, in tension as well as in compression (Figures 4.5 and 4.6). They have the advantage of being readily installed and removed for maintenance.
- Snubbers are standard, catalog supports that restrict dynamic movements without restricting its thermal expansion. There are hydraulic snubbers and mechanical snubbers (Figure 4.6).

Do nuclear power plants have more snubbers than other industrial facilities?

Yes. It is due to the more exacting seismic qualification requirements applied to nuclear power plant piping systems and the large seismic input in

Figure 4.5 Vertical struts mounted in tandem.

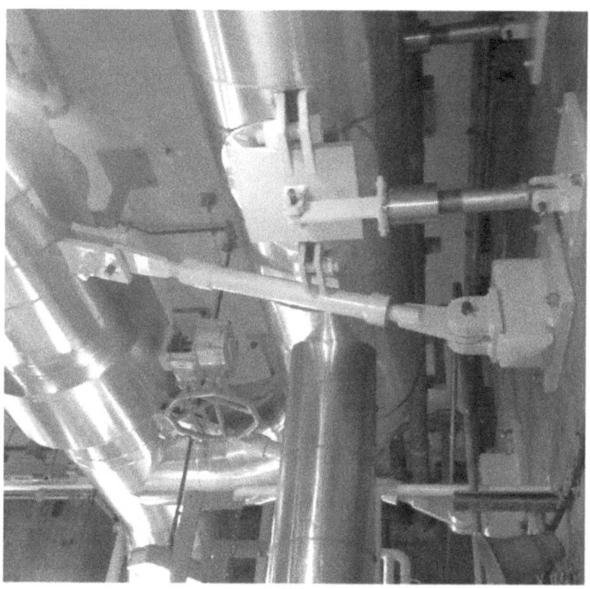

Figure 4.6 Lateral strut (below) and mechanical snubber (above).

the form of ISRS. But snubbers are active components that must periodically be inspected and maintained to verify that they will perform their seat belt function: expand–contract as the pipe thermally, slowly expands and contracts, but then lock when the pipe is subjected to dynamic motions such as those caused by a seismic event or water hammer. Snubbers are therefore periodically tested. When snubber malfunctions are discovered during surveillance testing, the snubbers are disassembled and inspected to understand the root cause of the problem. The inspection includes visual inspections as well as verifying the snubber piston movement, lock up and bleed, in both directions: extension and compression. Malfunctions of snubbers have been traced to damage by handling and installation, flow-induced vibration, or overload caused by a transient, usually a water hammer (IN 79-01). Wear and vibration have caused service-related degradation of hydraulic snubbers, including worn poppets, spring capture, and piston wear (IN 82-12). In a few cases, the malfunction of snubbers during surveillance testing was traced to a fabrication defect, such as solid particles left in the hydraulic fluid reservoirs, or inadequate tolerances of the snubber internals (IN 83-47, IN 84-67).

Any other precaution related to snubbers?

Two things come to mind. First is that the viscosity of hydraulic snubbers' fluid increases at large radiation exposures, in the range of 20 Mrad or more, and can reach incipient gelation. This type of exposure is possible in certain areas inside containment, in which case the fluids should be replaced before a fraction of this 20-Mrad radiation limit is attained (IN 80-42). Second is that snubber attachments can bind if the swing angle of the snubber rod is too large, and this binding will restrict the thermal movement of the line. This happened in one case and was resolved by modifying the snubber end connections to provide sufficient clearances to not bind as the pipe expands (IN 83-20).

Any special precaution when using catalog load-rated supports?

One must be careful with the load rating of rigid struts. This load rating depends on the rod material, and over the years several materials have been used by strut manufacturers, such as SA-36, SA-306 Grade 60, and A-108 Grade 1018, that have different mechanical properties. The manufacturer must make sure that the catalog load rating corresponds to the rod material supplied, and the designer must use the catalog value that is specific to the material installed (IN 79-10).

What are engineered support structures?

Engineered steel support structures are frames, baseplates, and plate and shell supports. Linear supports are steel members typically composed of hollow tube steel, angle iron, C-channels, I-beams, etc. Plate and shell members are simply support parts made of plate such as baseplates, curved sections, or intervening plates at flanged or bolted joints between members. These steel frames and their baseplates are custom-designed for the pipes to be supported and the space available.

How are support stresses analyzed?

They are designed and analyzed (1) by hand calculations for simple structures (simply supported beams or cantilevered or braced cantilever connections), (2) by special-purpose software, which includes ASME III NF or AISC postprocessing, or (3) by general-purpose FEA techniques (computer codes) for complex support structures or extreme loads such as caused by impact from pipe whip.

The analysis and qualification of pipe support members and joints pays close attention to each category of stress (tension, shear, compression, buckling) and how they are combined. Special consideration is given at the connections of members to one another and to the support connection to the building steel or concrete to ensure that the support model reflects the actual field conditions (pinned, fixed, guided, etc.). Localized effects between connecting support members are accounted for in the qualification of the support structures, for example:

- Punching shear when a steel member is welded perpendicular to a larger hollow tube steel. In this case an intermediate plate may be required at that connection to alleviate the effect of the punching shear load.
- Local stresses when two members connect close to each other, along a main support member.
- Web instability effects on open section structural members.

How are concrete anchor bolts qualified?

There are two general types of anchorage to concrete: cast-in-lace bolts and postinstalled (expansion) anchors. In either case, the support member is welded to a baseplate and the plate is secured to the concrete by anchor bolts. The plate may be embedded flush in a wall or may be bolted, protruding from the wall. The concrete anchor bolts in nuclear power plants are typically designed, analyzed, and qualified to ACI-349. For nonsafety-related anchor bolts, the design rules of ACI-318 have also been used.

Figure 4.7 Instrument tubing run along the wall.

How about supports for lighter-weight commodities?

Struts directly bolted to the wall or ceiling are commonly used for lightweight commodities such as instrument tubing and the instruments themselves, as in Figures 4.1, 4.7, and 4.8. Strut load limits, typically per unit length of strut, are provided by the strut manufacturer for nuclear applications.

What are the important considerations when working with postinstalled (expansion) anchors to concrete?

The important considerations, in addition to qualifying the anchor bolt group to the ACI code, relate to the bolt material and its installation. The NRC informed plants of one particular case of general breakdown in the installation of postinstalled anchor bolts at one plant. It started with the plant staff noticing loose postinstalled drilled anchors on recently installed pipe supports. The investigation of this condition pointed to a broader issue with the installation of concrete anchor bolts and resulted in the replacement or rework of thousands of anchor bolts. The problems recorded read like a cautionary tale (IN 10-01):

- Bore holes with the wrong depth
- Expander sleeves not driven deep enough
- Too few anchors for the load

Figure 4.8 Stainless-steel instrument tubing supported on strut tracks.

- Anchors installed in the wrong position
- Missing washers
- Wrong anchor types
- Abandoned drilled holes that were left as is, not plugged

Further discussions related to the installation of concrete anchor bolts can also be found in NUREG-1055, "Improving Quality and the Assurance of Quality in the Design and Construction of Nuclear Power Plants"; IN 07-04, "Construction Experience Related to the Assurance of Quality in the Construction of Nuclear Facilities"; Regulatory Guide 1.199, "Anchoring Components and Structural Supports in Concrete"; and Inspection and Enforcement Bulletin 79-02, "Pipe Support Base Plate Designs Using Concrete Expansion Anchor Bolts."

How about the case where the support is welded to the pipe?

This happens at anchors and in support arrangements that use trunions (cylindrical attachments) or lugs (rectangular attachments). The formulas for qualifying local stresses in welded attachments, hollow cylindrical trunions, or rectangular lugs are in ASME III Appendix Y.

What loads apply to pipe supports?

The design input for pipe supports consists of (1) the pipe support loads and movements, (2) the deadweight of the support and the seismic inertia load due to its own weight, labeled self-weight excitation, (3) the building differential movements at the support attachments points, and (4) friction loads.

Pipes supported by a trapeze configuration (dual snubbers, dual struts, or dual spring cans such as in Figure 4.4) may be subjected to a bending moment due to the eccentricity of the load or due to an uneven load distribution in the vertical direction. Remember that in a dual arrangement the two parallel elements (snubbers, struts) may not lock at the exact same instant, and therefore one element of the tandem may see the full load before the parallel one has time to lock and react part of the load. This has happened in plant operation and can be addressed by designing each element of the support tandem for the full load and accounting for the offset of the support reaction load on the pipe.

Regarding pipe-to-support frictions, it is typically not included in the piping analysis model; the piping analysis output is based on frictionless supports. Instead, the support designer studies the thermal motion of the pipe and applies a thermal motion-induced friction load in the resisting direction, opposite to the direction of thermal motion.

How are anchor loads calculated?

Terminal anchors are designed for the total loads from both sides of the anchor, in other words from two different piping analyzes. Where the anchor is an interface between a seismic piping system and a nonseismic piping system, then we have seismic loads from only the seismic side of the anchor. In this case, either a seismic load from the nonseismically designed pipe side must be conservatively estimated or the anchor may need to be designed to sustain plastic moment loads from the nonseismic side in order to protect the seismic side in case of an OBE or SSE. Designing for plastic moments in the pipe has resulted in massive anchors.

What are the fabrication and construction requirements for piping systems and supports?

There are basically four essential aspects to shop fabrication and field construction:

- The qualification of processes and personnel involved in fabrication and construction
- The compliance of the fabrication and construction process to qualified procedures
- The nondestructive examination (quality control NDE) of the constructed assembly
- The testing following completion of examinations, which consists of (1) leak or pressure testing and (2) functional testing before turnover to operations

ASME III Subsections NX-4000 for piping and NF for supports address these requirements, except for the functional testing, which is addressed in ASME XI for pressure boundary and ASME QME for snubbers and active components. Each project develops plant-specific shop fabrication, procurement, field construction, examination, and testing procedures and drawings to meet these requirements. The installation of the piping system and its supports then strictly follows these fabrication and construction procedures and drawings. Deviations must be carefully reconciled and documented by Engineering.

How about flange joints?

Welding of the piping systems is well managed through a combination of the ASME IX welding program and weld inspections and NDE during fabrication and erection. Mechanical fabrication, such as bending, and mechanical joints, such as threading, swage fitting, and bolting, should receive a similar level of rigor as welding. Qualified mechanical joining procedures must be followed, for example, tubing fittings should be installed or retightened using the manufacturer installation instructions and code requirements, in addition to design-specific requirements.

Good practices for bolting of flange joints in pressure equipment (piping, valves, pumps, tanks, and vessels) are addressed in ASME PCC-1 "Guidelines for Pressure Boundary Bolted Flange Joint Assembly." While this is a general standard, developed with the fossil power, chemical, and hydrocarbon industries in mind, it provides good practical engineering and maintenance guidance for nuclear power plant engineers.

What are the requirements for pipe bending?

Regarding pipe bending in the shop, the modalities of bending are addressed in ASME III NB/NC/ND-4000. Following the bending operation, each bend is typically inspected for bend thickness, ovality, absence of wrinkles, and surface finish. Cold bending of stainless-steel pipe can be performed using qualified bending machines operated by experienced operators, starting with a fully annealed pipe, and achieving a bend radius no less than 3 pipe diameters. Bending to tighter diameters, as small as 1.5 pipe diameters, is attainable with other qualified means such as induction bending.

How about the surface appearance of the constructed assembly?

Workmanship is a quality that the nuclear power industry takes seriously. As we walk around the plant, we notice the cleanliness and the "like-new" look of SSCs, which are regularly inspected, maintained, repaired, cleaned, coated, and painted. In particular, piping system must be free from surface damage such as scratches, cracks, dents, rust. A good reference for workmanship and surface appearance can be found in the ASTM pipe specifications themselves.

4.2 TUBING SYSTEMS

Is there a difference between tubing and piping?

Piping is fabricated to a commercial size in accordance with ASME B16.10 or B16.19 and is therefore procured to a pipe schedule called out in these two B16 standards. Tubing could be defined as a conduit for fluids, fabricated to any size other than a B16 schedule. Tubing can be used to convey a fluid (like a pipe) or in bundles in heat exchangers. Nuclear power plant tubing is generally used in small sizes for instrumentation, but tubing can be procured in large sizes. For example, copper tubing can be procured all the way up to 12-inch diameter. Finally, unlike piping, tubing does not have to be cylindrical.

How is the tubing size specified?

The simple answer is to refer back to the tubing specification, either ASME II or ASTM. For example, ASTM A632 "Standard Specification for Seamless and Welded Austenitic Steel Tubing (Small Diameter) for General Service" will refer back to ASTM A1016 for "General Requirements." In turn, ASTM A1016 requires the buyer to specify the outside diameter and average or minimum wall thickness required for each order. Tolerances on thickness are provided in either the source specification (ASTM A632 in this case) or the general specification (ASTM A1016 in this case). The tubing specification requires a flaring test, since tubes can be joined by flare connections.

Another common example of tubing is copper tubing, ASTM B75 "Seamless Copper Tube." In this case, the dimensions of copper tubes are specified in ASTM B251. For ASTM B88 "Seamless Copper Water Tube," the tube dimensions are provided in a table in the specification itself, from 1/4 up to 12 inches.

Finally, since we mentioned piping and tubing, we could also mention what is called line pipe; this is a high-quality and high-strength carbon steel pipe procured to API 5L for use in the oil and gas pipeline industry, and is also found in refineries and petrochemical plants. The size of line pipe is in accordance with size tables in API 5L.

How is tubing used in a nuclear power plant?

In nuclear plant applications, tubing systems are used primarily as air supply to operate and control components, such as air-operated valves, or supply to instrument impulse lines and sensing lines, or to measure a process variable such as temperature, pressure, or flow.

What is the safety class of tubing systems?

The safety classification of a tubing system is, like piping, set by the function they perform. Therefore, tubing systems can be either safety or nonsafety, seismic or nonseismic, essential or nonessential, depending on the function they perform or on the consequence of their leakage or rupture.

Is instrument tubing, like piping, designed and installed to the ASME code?

Yes, if instrument tubing is safety related, it is designed and fabricated to ASME III Class 1, 2, or 3 (or B31.1 for non-ASME III plants). If it is nonsafety related, it is designed to ASME B31.1.

How do tubing systems differ from piping?

As we have seen, they come in different sizes than piping. In nuclear plants, for instrument and control applications they are almost always made of stainless steel. They are cold bent in the shop or in the field, typically to no less than a three-diameter radius, rather than welded to elbows like pipes. Accuracy in using benders is a key attribute of a tubing fitter. The tubes are then joined by mechanical means, most often by swaged fittings rather than by welding, which permits these joints to be disconnected for maintenance and instrument calibration and checks.

Good workmanship, by qualified tube benders and fitters, is important to achieve leak-free construction. The tube must enter straight into the fitting and be correctly torqued; otherwise the swaged joints can leak. Leaks are often traced to breakdowns in workmanship, particularly inadequate insertion and incorrect torquing of the fitting to the tubing. Also, because tubing is joined by swaging, their surface hardness, their fully annealed heat treatment, and their surface finish are important to achieve a leak-free joint. Tubing fitting manufacturers provide valuable information and training to design engineers, construction engineers, and maintenance technicians on how to correctly lay out and assemble tubing.

Any special precaution regarding the means by which tubing is supported?

Because tubing has a small size and a thin wall, it must be routed as much as possible close to walls so as to be protected from accidental damage, as shown in Figures 4.7 and 4.8. Sometimes tubing is run in trays for protection, as well as to prevent sagging (Figure 4.9). Tubing is run in groups, in parallel layout as much as possible (Figure 4.10). In particular, capillary tubing is run in tube tracks or inside struts or channels for protection.

Figure 4.9 Tubing run in a tray.

Figure 4.10 Tubing routed and supported in a consistent manner.

During maintenance near tubing, temporary protective barriers may be erected to prevent accidental impact.

How important is the slope of instrument tubing?

Tubing slope is an essential and exacting design feature for the correct operation of the instruments and controls served by the tube. The slope of instrument lines is a critical parameter. Liquid-sensing lines must slope downward from the process connection to the instrument to prevent air entrapment. Gas-sensing lines must slope upward from the process connection to the instrument to prevent entrained condensate to flow to the instrument. Layout guidance is provided in the International Society of Automation (ISA), formerly the Instrument Society of America, ANSI/ISA-67.02.01, "Nuclear Safety-Related Instrument-Sensing Line Piping and Tubing Standard for Use in Nuclear Power Plants"; NRC Regulatory Guide 1.151, "Instrument Sensing Lines"; and ANSI/ISA-S67.10, "Sample-Line Piping and Tubing Standard for Use in Nuclear Power Plants."

How are tubing systems analyzed and qualified?

ASME III and B31.1 do not differentiate between piping and tubing systems. Tubing systems are analyzed and qualified in the same way as piping systems, except that because of their light weight and small size (typically below 2 inch), they are often routed and supported to prequalified span configurations using so-called cookbooks. This makes for uniformly and evenly laid-out banks of tubing (Figure 4.10). Expansion loops made by bending the tube are used to absorb thermal expansion in hot tubing systems.

If tubing systems are essential for shutdown following a given pipe break, they must be protected from pipe whip and jet impingement from that break.

What are the key steps in developing prequalified tubing spans?

First step: Assemble the tubing information: its material specification, its physical properties (modulus of elasticity and coefficient of thermal expansion), its size (diameter, wall thickness, and section modulus), its linear weight, and the weight of in-line components.

Second step: Determine the design pressure and temperature, and with the material and design temperature obtain the allowable stresses at ambient and hot temperature from ASME II Part D.

Third step: Determine the loads and load combinations for each service level. For tubing systems the typical load combinations are similar to piping.

Fourth step: Determine the peak OBE and SSE spectral accelerations at the envelope elevation of the attachment points, in each of the three directions.

Fifth step: Treat a span of tubing between supports as a beam on supports, apply 1.5 times the resultant lateral peak OBE and SSE accelerations, and obtain the maximum stress as a function of the span length L, for example, $\sigma_L \approx i\,g\,w\,L^2/(10\,Z)$, where σ_L is the longitudinal stress caused by the OBE or SSE; i is the stress intensification factor, which some (not all) tubing fittings manufacturers will have; g is the applied lateral OBE or SSE acceleration (including the weight and the 1.5 amplification factor); w is the linear weight; L is the span length; and Z is the section modulus. Limit the span stress to a fraction of the ASME III or B31.1 allowable stress, for example, half the allowable stress for the service level minus the longitudinal stress caused by the internal pressure. Writing $\sigma_L < k\,S_h - (PD/4t)$, we can solve for the maximum span length, L, permitted. In this formula, k is the fraction of the allowable stress in accordance with the FSAR, S_h is the hot allowable stress, P is the maximum internal pressure, and D and t are the tube diameter and thickness, respectively. Refinements are necessary when there are in-line components and a further reduction for bends and elbows. The reaction load at the ends of the beam model constitutes the support loads.

Final step: The maximum span (maximum distance permitted between tubing supports) is used by the plant design group to route the tubing system. The final route isometric (Figure 4.11) is then compared with the standard span criteria at the design stage and at the as-built stage; to confirm compliance, standard supports are selected, and the support loads are used to size the supports and verify the backup structure.

How are tubing supports designed and qualified?

Safety-related tubing supports are typically designed and qualified to ASME NF, and nonsafety supports to the AISC "Manual of Steel Construction" with welding to AWS D1.1 and D1.3 "Structural Welding Code." Because the support loads caused by tubing are small, many times tubing is supported by supports made of bolted strut assemblies (Figures 4.7 and 4.8),

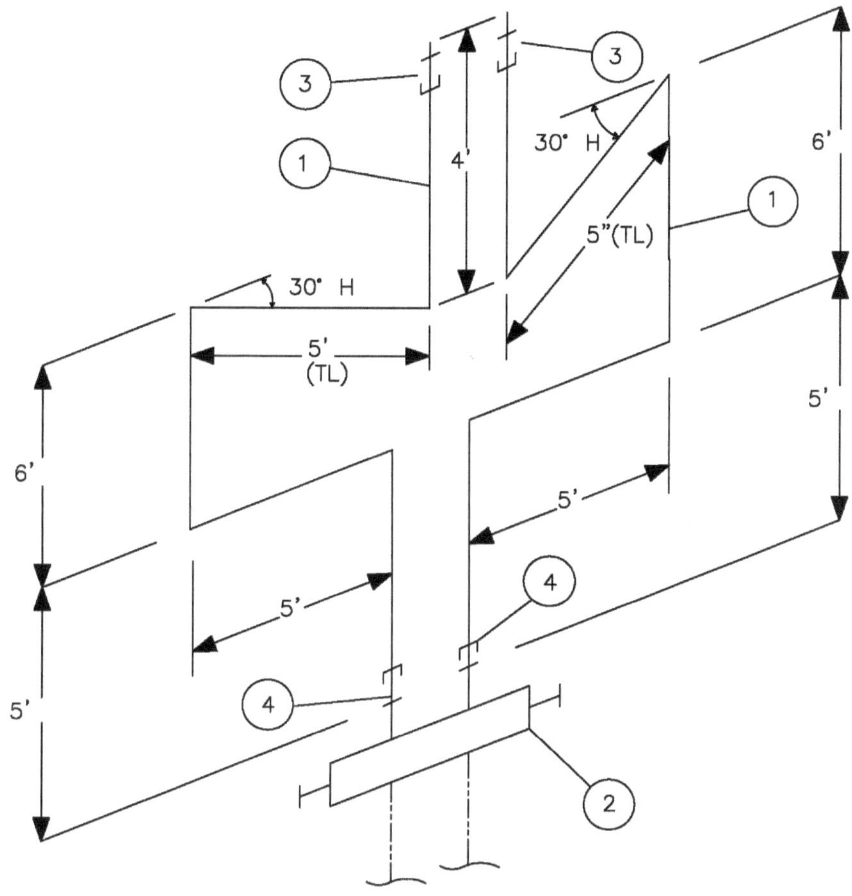

Figure 4.11 Example of field isometric to verify instrument tubing spans.

which saves installation time and reduces congestion when compared to supports made of welded structural steel.

Is the construction of tubing systems also in accordance with ASME III?

Yes, ASME III has provisions for mechanical joints commonly used with tubing systems. The installation is then examined, typically visually, and leak tested to ASME III or to ASME XI requirements if it is a joint remake following a maintenance or a modification. When instrument tubing systems are hydrostatically tested, they should be tested with the same liquid as the process lines. Tubing systems are preferably tested pneumatically.

4.3 HEATING, VENTILATION AND AIR CONDITIONING DUCT SYSTEMS

What constitutes a heating, ventilation and air conditioning system?

Heating, ventilation, and air conditioning (HVAC) systems start with the supply and exhaust fans and the air handling units. The air is distributed through the ductwork (Figures 4.12 and 4.13), which includes louvers and insulation. Duct accessories include balancing dampers, extractors, access doors, flexible connectors, grilles, registers, diffusers, etc. Air handling units include dampers, plenums, air cleanup and filter units, coil units, thermostats, ventilators, etc. Also, the HVAC system includes the refrigerant systems, air flow measuring devices, instrumentation and control, penthouses, demisters, silencers, vent stacks, and duct accessories that conform to their own purchase specifications.

What is the function of HVAC systems in a nuclear power plant?

The HVAC systems provide building and rooms cooling so that equipment remains functional, in summer and winter environments, and after postulated breaks that will discharge hot steam from leaks or breaks. For safety-related equipment, cooling must be provided even in case of accidents that could cause large environmental temperatures in rooms or

Figure 4.12 Heating, ventilation, and air conditioning duct and duct support vertical riser.

compartments, in places exceeding 200 °F. For example, the containment spray (CS) system is normally on standby at room temperature; however, during accident conditions, the line temperature can exceed 200 °F, and therefore the CS pump room temperature would rise. In this case, the HVAC system would be relied upon to perform a safety function by keeping the CS pump room sufficiently cool so that the CS pumps can run and provide CS.

Another essential function of the HVAC system in a nuclear plant is to provide the filtration functions necessary to confine accidental releases and contamination.

What criteria apply for the structural design of the ductwork support system?

The performance requirements of nuclear safety-related HVAC systems are addressed in the ASME Code on Nuclear Air and Gas Treatment (AG-1), including the structural design requirements for safety-related, Seismic Category I HVAC ducts, with stress limits in AG-1 SA-4220, and deflection criteria in SA-2430.

Safety-related duct supports can be analyzed and qualified in accordance with ANSI/AISC N690. Cold-formed members are qualified for seismic-induced fatigue cycles in accordance with the American Iron

Figure 4.13 Heating, ventilation, and air conditioning duct and duct support horizontal run.

and Steel Institute (AISI) "Specification for the Design of Cold-Formed Steel Structural Members." Mechanical joints in ducts and in connections of support members have been seismically qualified by testing.

Non-safety duct supports are normally designed to the requirements of the AISC "Manual of Steel Construction." The HVAC ducts are designed in accordance with the applicable provisions of ERDA 76-21 "Energy Research and Development Administration - Nuclear Air Cleaning Handbook, Design, Construction and Testing of High Efficiency Air Cleaning Systems for Nuclear Application" and SMACNA "Standards for Low or High Pressure, Rectangular or Round Duct." Welding is typically in accordance with AWS D1.1 "Structural Welding Code," with AWS minimum weld size provisions.

Many duct supports are multitier. Where possible they are made of strut assemblies bolted together (Figure 4.14) rather than welded angles (Figure 4.13); this eliminates the need for field welding the support members and provides valuable flexibility to add or modify the support structure.

How are duct systems structurally analyzed and qualified?

First, the design loads and load combinations are defined for the project. Typically, the loads include the weight of the system, with a minimum concentrated live load of 250 lb, the pressure in the duct (differential, negative, and positive), the thermal growth (including postaccident hot environments, if applicable), and the seismic loads. The duct is protected from missiles or high-energy line break effects, rather than analyzed for these loads that can typically puncture the duct. ASME AG-1 SA-4211

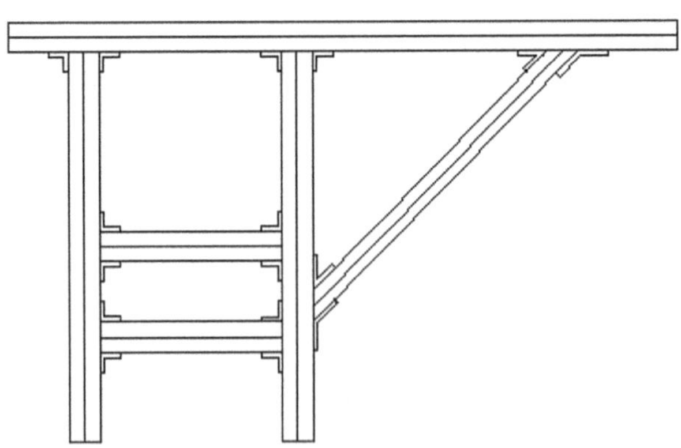

Figure 4.14 Cable tray support with lateral brace using struts.

also addresses momentum loads caused by flow. Finally, if the duct is outdoor, it is designed for wind and, if safety-related, for tornado loads and missiles.

Then, the duct is modeled and the loads are applied to the model. This can be done by a three-dimensional computer model of the duct runs and their supports, or by analyzing bounding spans of ductwork. The seismic analysis of the duct system uses beam approximations, with the correct weight distributions and offsets, and the correct natural frequency, rather than a shell elements model.

The seismic analysis can be static, with lateral and axial static acceleration, or dynamic, typically by modal analysis. The static analysis provides reasonable results, owing to the generally simple layout of the ductwork. However, if there are complex geometries or changes of duct size that can cause strain concentration in smaller ducts attached to larger ducts, then a dynamic analysis may be necessary. The damping value for ductwork is provided in Regulatory Guide 1.61. It depends on the type of duct joints. Unlike piping systems, ductwork does not contain anchors to separate duct runs into separate analyzes. It is therefore necessary in some cases to model sections of ducts with an overlap on adjacent spans. If the model is terminated in an overlap, two additional duct spans in the overlap region can be included to simulate the boundary conditions.

Once the beam-based results are obtained from the duct system analysis, the loads are converted to local stresses and compared with limits to prevent sheet metal tearing or crimping, prevent buckling of the stiffeners, prevent opening of the duct joints, and qualify the supports and their attachments to the duct and to the backup structure. If there are in-line active components, such as dampers, they must also be qualified for operability and/or leak tightness.

4.4 CONDUIT SYSTEMS

What constitutes a conduit system?

A conduit system consists of the conduits themselves, and also junction boxes (Figure 4.15) couplers and unions, and their supports. Junction boxes are enclosures that provide a degree of protection against incidental contact with the enclosed cables and equipment, and provide a degree of protection for the enclosed cables and equipment against specified environmental conditions.

Figure 4.15 Conduit and junction box.

What is the function of conduits?

Conduits convey electrical cables to and from termination points at equipment, and to cable trays, in a manner that protects the cable from damage. Conduits and cable trays together are often referred to as raceways.

How are conduit systems installed?

Conduits are normally installed by coupling together individual segments using threaded couplers or unions. So-called heavy couplings are used in 90°turns. Some couplings are designed to allow for cable pulling and for connecting flexible and rigid conduits. Conduit runs, composed of the coupled segments, have termination points that can be drops of cable to and from cable trays, or flexible connections to equipment (Figure 4.16) or junction boxes (Figure 4.15). The junction boxes can be supported separately or they are supported by the rigid conduit itself.

How are conduit systems structurally qualified?

The conduit systems and their support must be structurally designed and qualified to remain functional for normal operating conditions, and in the

Figure 4.16 Conduit support structural steel and struts.

case of safety-related cables, they must be designed to remain functional after accident conditions such as the hot and humid environment caused by a postulated pipe break, and the inertial and anchor motion loads caused by an earthquake. Safety-related conduit systems are normally protected from the effects of pipe rupture by separation and from tornado by enclosures.

Conduits are typically qualified by spans of standard configurations. The equivalent static method is developed based on the ISRS and accounting for the conduit system frequency. Based on whether the actual span frequency is above or below the peak spectral frequency, a multimode amplification factor of 1.5 or a lower value is used to develop the conduit support loads. Because there are different ISRS in the buildings, there will also be different loads depending on the location of the conduit in the building. Standard bounding support loads are determined based on the conduit size and the span length.

For multiple conduit runs of different diameters on common supports, the conduits with the most stringent criteria normally govern the spans and the loads are added to design the common gang support.

Are there allowable stresses on conduits, like there are on piping, tubing, or ducts?

Allowable stresses for conduit and junction boxes are normally in accordance with AISC "Specification for the Design, Fabrication and Erection of Structural Steel for Building" and the National Electrical Manufacturers Association (NEMA) Code or manufacturer's limits. Welds are usually designed and qualified in accordance with AWS D1.1, "Structural Welding Code."

How are the junction boxes qualified?

The junction box itself may be seismically qualified by testing, accounting for the load applied by the conduit in all the possible orientations of the box. The design of the junction box support considers the weight of the junction box and its content, the conduit, and the support members. Calculated forces or stresses are compared with test-based manufacturer's limits.

How are the conduit supports qualified?

Struts with clamps (Figure 4.8) or structural steel with *U*-bolts (Figures 4.15 and 4.16) are used to support and restrain conduit systems. Struts are preferred where possible because they are more readily constructed and modified for maintenance. For strut support members, the actual allowable loads are normally determined by testing or by specific load capacities in accordance with the AISI "Cold-Formed Steel Design Manual." Where intervening structures are required, then custom-designed steel support structures are used (Figure 4.16).

If the analyzes reveal that torsional moments are capable of generating warping stresses in members with open cross-sections, these warping stresses are to be added to the calculated normal and shear stresses. Various eccentricities must be considered to realistically account for the application of loads and connections between structural members.

How to address conduit embedded in concrete?

The concrete is considered a multidirectional support. The bond stresses between conduit and concrete or grout should be verified against an allowable bond stresses that can be calculated, for example, in accordance with the ACI-318 Code. Thermal effects are also considered in the design of embedded conduit.

4.5 CABLE TRAYS SYSTEMS

What is the function of cable tray systems?

Cable trays distribute bundles of electrical cables from power supplies to electrical equipment and components throughout the plant. Cable tray systems consist of insulated electrical cables layered inside metallic trays, which are supported from concrete walls and ceilings, or steel structures (Figures 4.17 and 4.18). Most trays in nuclear power plants are of the

Figure 4.17 Cable tray to wall penetrations.

Figure 4.18 Seismic-supported cable trays.

ventilated rail type, with openings and cutouts to permit connections, allow for cooling, and allow for draining of water that may accidentally spill on a tray.

Cable trays in nuclear power plants are most often made of steel (galvanized steel or stainless steel). The cable spans consist of straight runs and fittings (bends, risers, etc.). They are often supported by trapeze-configured supports with the cables pulled through the tray (Figure 4.18). The trays would be labeled in accordance with their safety designation for ease and accuracy of identification in the field. This is important to ensure that work is done on the right run of cable, or if the cables are inside a conduit, then only the affected cable is de-energized and worked. It is important, if a cable is to be cut, that it should first be clearly identified and positively tested as de-energized.

In some cases, cable trays are supported and braced using strut systems; this eliminates the need for welding (except to the backup structure or baseplate) and permits assembly–disassembly of the trays for maintenance.

How are cable trays classified as safety related?

Oftentimes, individual safety-related cables are difficult to identify among all other cables in a cable tray. In this case, the complete bundle of cables in the tray containing one or more safety-related cables is classified as safety related.

What criteria apply to the design of cable trays?

Cable trays are designed and installed to NEMA standards such as NEMA VE-1 for design and NEMA VE-2 for installation. The National Electric Code then dictates the requirements and limitations for cable loading. In addition, safety-related cable trays or cable trays that can constitute a seismic interaction are seismically designed. Essential cable trays must also be protected from direct missile strikes and from pipe whip or jet impingement impact loads from postulated high-energy line breaks.

The important considerations for cable trays are their resistance to fire, the potential for ignition and propagation of cable fire between adjacent trays. This is related to the cable materials, the layout of the cable trays, the fire protection system, and the area ventilation system. Cable trays at risk of fire are sprayed with fire retardant. In one classic case, technicians were checking the direction of air flow through a cable wall penetration using the flame of a candle. But the penetration had been filled with a temporary seal that was combustible. The temporary penetration seal caught fire, which spread to the reactor building, causing significant damage to the cables (BL 75-04).

How are cable trays seismically supported?

Seismic category SC-I or SC-II/I (source of interaction) cable trays are braced laterally and longitudinally to resist seismic (OBE and SSE) loads (Figure 4.18). The bracing system is typically designed by static seismic analysis applying a load based on the peak spectral acceleration to each tray span in each of the three directions. The damping for cable trays is defined in NRC R.G. 1.61 and depends on the magnitude of the earthquake (OBE or SSE), the presence of fire retardant, and the cable loading (near-empty tray or full tray). Commercial, nonseismic cable trays are typically supported on 15–25 ft intervals. In the case of nuclear power plant cable trays, it is the seismic analysis that will govern the cable tray lateral and longitudinal bracing.

The seismic design of cable trays in nuclear power plants benefits from extensive investigations of their behavior and performance in real earthquakes and in tests. Key documents addressing this subject include EPRI NP-7149-D "Summary of the Seismic Adequacy of Twenty Classes of Equipment Required for the Safe Shutdown of Nuclear Plants," EPRI NP-7151-D "Cable Tray and Conduit System Seismic Evaluation Guidelines," and EPRI NP-7152-D "Seismic Evaluation of Rod Hanger Supports for Electrical Raceway Systems."

CHAPTER 5

Supporting Structures

Chapter 5 describes the various types of steel and concrete structures that serve to support systems and equipment. We review their design basis and their multiple uses when the plant is in operation. We address the use of scaffold in nuclear power plants, as temporary or permanent installations, and the cautions that apply when the scaffold is permanent. This chapter also discusses the question of clearance between structures and commodities, a point of particular interest in the congested rooms and compartments of a nuclear power plant.

5.1 STRUCTURAL STEEL

What are the different types of steel structures in a nuclear power plant?

The different types of steel structures in a nuclear power plant include, roughly from the largest to the smallest,

- steel that is part of the building structure (beams, columns, etc.)
- structural steel frames inside building compartments and rooms
- steel floors and platform framing
- monorail and crane support structures
- tornado missile shields
- jet and whip restraints
- supports of suspended and distribution systems
- baseplates
- stairs, handrails, ladders, gratings and checkered plates, gates, doors, and security barriers

What is the design function of these steel structures?

Steel structures act primarily as supports, restraints, and reinforcement. They are designed for normal dead loads, live loads, thermal loads, and accident loads, which are typically orders of magnitude larger than the normal loads. Safety-related steel structures are designed for a direct fire or they are protected by fireproofing.

Nuclear Power Plant Safety and Mechanical Integrity
ISBN 978-0-12-417248-7

What are the design codes for support steel structures?

This depends on the vintage of the plant. For the more recent plants, the design codes are as follows:

- ASME III Div.1 NF for supports of ASME III systems and components
- AISC N-690 "Specification for the Design, Fabrication and Erection of Steel Safety-related Structures for Nuclear Facilities" for whip restraints and for structures designed to resist impulsive loads such as missiles
- AISC "Manual for Steel Construction" for non-ASME NF structural steel
- AISI "Specification for the Design of Cold-Formed Steel Structural Members"
- ACI-349 "Code Requirements for Nuclear Safety-related Structures" for safety-related concrete and anchor bolts
- ACI-318 "Building Code Requirements for Reinforced Concrete" for nonsafety-related concrete and anchor bolts
- ACI "Detailing Manual"
- AWS D 1.1 "Structural Welding Code"
- AWS D 1.4 "Reinforcing Steel Welding Code"
- ACI-359 and ASME III Division 2 for concrete containments
- ASME III Class MC for metal containments

Within AISC, the following standards are commonly used:

- AISC S 335 "Specification for Structural Steel Buildings, Allowable Stress Design and Plastic Design"
- AISC 341 "Seismic Provisions for Structural Steel Buildings"
- AISC "Load and Resistance Factor Design Specification for Structural Steel Buildings"
- "LRFD Specification for Steel Hollow Structural Sections"
- "LRFD Specification for Single Angle Members"

How are steel structures analyzed and qualified?

In today's designs, the building and steel structures are modeled and analyzed by finite element methods. This was not the case in the 1960s and early 1970s, when much of these structures were analyzed by hand calculations of steel trusses. The static and dynamic loads are applied to the model, and reaction loads (forces and moments) are calculated in members and joints of the model. The resulting stresses and deformations are compared with codes and design specification (D.Spec.) limits. The level of detail of the model depends on the vintage of the plant. It varies from stick models, all the way to detailed finite element analysis (FEA) models, with soil–structure interaction for the building structures and floor flexibility. The equipment and system support structures, the ASME III NF structures, and the N-690 whip restraints are modeled and analyzed individually.

The applied loads are, first, the normal deadweight and live loads. For slabs, the member's weight corresponds to the sum of the deadweight of steel and reinforced concrete members, normally added to the specified live load, typically up to 100 psf, plus a load allowance for the systems and components, which may be in the order of an additional 50 psf, unless there is uniquely large equipment mounted on the floor.

After the design is verified for normal loads, it is analyzed for postulated accident. The seismic load affects all the steel frames, concurrently. The pipe break loads consist of possible whip impact, jet impingement, and the loads caused by the thermal growth of the steel frames, constrained by the smaller growth of the reinforced concrete, under large postbreak ambient temperatures that can reach 300 °F, depending on the location inside containment. The whip impact loads may be the largest of all, and the pipe break loads have to be combined with the safe shutdown earthquake (SSE). Outdoor structures have to be designed for high winds and, for safety-related targets, for tornado missiles. Inside and outside steel structures have to also be designed for missiles generated by postulated broken and ejected parts and components.

Is the design of structural steel structures always elastic?

For structural steel, the required section strength for normal operating loads is based on the elastic design method. The elastic stress limits are specified in standards such as AISC Code or ASME III NF. The section strength required to resist large postulated accident loads can be based on plastic design methods described in Part 2 of the AISC Code, or ASME III Appendix F for Faulted, Service Level D loads.

How about scaffolds, are they considered steel structures?

Yes, scaffolds are steel structures. They receive special attention because of personnel safety and because of the potential for adverse seismic interactions with nearby safety-related structures, systems, and components (SSCs). Scaffolds have their own special rules and will be discussed later in more detail. They have to meet the installation standards in 29 CFR Part 1926, Subpart L "Scaffolds," Section 1926.451(a)(1) "General Requirements for Capacity," and Appendix E as well as the guidelines of the OSHA 3150

Information Booklet "Scaffolds Use in the Construction Industry." Usually, plant procedures require permanent scaffold poles in safety-related buildings to be seismically designed to AISC. Such installations end up to be similar to other permanent structural steel and platforms found inside safety-related buildings.

Is this same level of design detail also applicable to commercial-type commodities such as handrails, ladders, and grating?

Yes, because they could be a source of adverse seismic interaction with nearby safety-related SSCs. They too have dual requirements: strength requirements and OSHA requirements. Grating and checkered plates are often designed in accordance with ANSI/NAAMM design practices and design criteria. In addition, they are designed for interaction loads. For example, it is usual for grating to be the target of whip and water or steam jets from postulated pipe breaks. In this case, the design has to assure they will not be dislodged and ejected becoming a secondary missile.

Are temporary rigging beams also designed in detail?

Definitely yes. Rigging can be for lightweight components such as valves, or for large equipment and structures, in which case it is referred to as a heavy lift and is a critical operation that must be planned and engineered in detail to prevent accidents. Rigging requires detailed lift plans; lift drawings; lift manuals; procedures for crane load test and witness load test; crane inspections; planning for crane erection and operation and demobilization activities; engineering analyses for lift loads, lift margins, wind load analysis on suspended objects with large sail areas to evaluate the potential for crane boom side load; crane foundation designs for safe load-bearing capacity; and design of lift lugs, spreader beams, and other special lift devices.

What are the keys to designing lifting devices?

The design of lifting devices must conform to ASME B30 series of standards, including the common B31.20 for "below-the-hook lifting devices." Typically, a minimum safety factor of 1.8 is recommended for lifting attachment design to conform to ASME B30.20. The design of lifting lugs must also conform to AISC Part 5. In addition to the lifting device itself, a minimum safety factor of 1.25 is used in analyzing vessel shells and skirts for lifting forces. In some cases, engineering has generically qualified existing supports, such as struts or beam, for lifting smaller loads, such as loads not exceeding 1000 lb.

When is an ASME III D.Spec. needed for steel structures?

A D.Spec. is needed for steel structures that are in the scope of ASME III Div.1 Section NF, which means ASME III pipe and mechanical equipment supports.

What should the D.Spec. cover?

The contents of a D.Spec. are outlined in ASME III Div.1 NCA. Practically speaking, here is what one would expect to find in an ASME III D.Spec.:

- The cover page PE stamped and signed, with a standard statement that the PE "to the best of my knowledge and belief...the Specification is in accordance with ASME III Division 1, Edition and addenda to be stated, Subsection NCA-3250..."
- The scope, which could be a single D.Spec. for all Class 2 and 3 piping and tubing supports, for example, with diagrams of the boundary between the NF support and the steel structures. For this purpose, the diagrams in NF are the best reference.
- How the supports are classified as safety class 1, 2, or 3, and as seismic category SC-I or SC-II/I.
- A list of project references and codes and standards.
- Responsibilities of the designer, the owner, the owner's designee, the supplier.
- Interfaces between the parties, ideally with a diagram of who provides what to who, and even interfaces within the design organization: Who develops the layout, who spots the supports, who updates the plant 3D model if there is one, who prepares the pipe isometrics, who prepares the support drawings and calculations, who provides loads and movements at supports, who calculates the support tributary weight and stiffness and when they are needed, who establishes the support insulation requirements and temperature from hot pipe to backup structure (for example, 100 °F/in along the support members, from the hot pipe surface outward), who provides the seismic response spectra at the right damping, who provides the stiffness of the backup structures, who selects the catalogs for standard catalog items (rods, U-bolts, beam attachments, etc.). Then what output the support group provides to other groups.

- Requirements for a design report: This is the report prepared at the end of the design, and which documents that the design requirements have been met and the support structures are qualified.
- Material requirements, for example, when to use stainless steel, how to treat the contact between stainless steel pipe or tubing and carbon steel supports, what nonmetallic materials are permitted or not permitted, toughness testing, carbon equivalent, weld filler materials, where is autogenous welding permitted, special heat treatments, etc.
- The normal and abnormal environments by building (or by room within the building for radiation and temperature), including gamma and beta radiation, postbreak chemistry and temperature (for example, 50–120 °F in normal operation inside containment, and possibly 350 °F for a short period after a pipe break accident), humidity, pressure (for example, −0.5 in water to +1 psig in normal operation inside containment, and maybe over 50 psi post-pipe break accident), etc.
- Loads and load combinations for each service level, and for design and test conditions.
- Applicable sections from ASME III NF-3000 for stress limits and other limits for each service level. Also, the D.Spec. would address whether Service Level C limits will be imposed on Level D loads to provide for the so-called functional capability of active components.
- Supplementary requirements beyond ASME III NF such as R.G. 1.124 "Service Limits and Loading Combinations for Class 1 Linear-Type Supports."
- Limits other than stresses, for example, deflection limits or stiffness requirements.
- Requirements for the design of in-line anchors, where loads will be needed from both sides of the anchor.
- Requirements on support gaps, and tolerances for all members, components, and anchor bolts.
- Requirements for bolted joints.
- Welding requirements, welding techniques, standard or minimum weld sizes, weld surface finish.
- Possibly a section on snubbers because they are unique active supports.
- Congestion and access requirements for in-service inspections.
- Quality assurance requirements.
- Even though this is a design specification, the question of as-built may be addressed, possibly within the context of roles and responsibilities.

5.2 SUPPORTING CONCRETE

What is the design function of concrete structures, liners, and embedded plates?

- Reinforced concrete structures are the steel bar (rebar)-reinforced concrete walls, domes, ceilings, and floors that constitute the building structure and its walls.
- Liners consist of welded steel plates attached to some of the concrete walls to protect the wall and provide leak tightness. The containment liner on the inside of the concrete containment building in a pressurized water reactor (PWR) is meant to maintain the containment leak tightness under normal operation and accidental pressurized release. The liner is the third containment barrier after the nuclear fuel cladding and the reactor coolant system.
- Baseplates are steel strips, sheet plates, or corner plates that are embedded or bolted to the concrete. They provide a face for welding or bolting to other steel structures and supports to the walls (Figure 5.1). The function of all these embedded or bolted plates is to provide a structural load path from the supported SSC to the reinforced concrete.

Figure 5.1 Bolted baseplate supporting structural steel.

What are the key structures in a containment?

The nuclear island of a PWR consists of a vertical cylindrical shell supporting a hemispherical dome, with the cylindrical shell being supported on a circular foundation mat and a reactor cavity pit at the center. The containment structure above the base mat is independent of the adjacent interior and exterior structures. Sufficient space is provided between the containment and the adjacent structures to prevent impact between structures under seismic design loads.

How is the liner anchored to the concrete?

The PWR containment structure is steel-lined. The entire inside face of the containment (the mat, the wall, and the dome) is lined with a continuous welded steel liner for leak tightness in case of a rupture inside containment. The liner is anchored to the reinforced concrete normally through studs welded to the plate that become engulfed with concrete at pouring (Figure 5.2).

The principal reinforcing steel used in the mat, the cylinder wall, and the dome are heavy duty rebars, over 2 in in diameter (No.18 rebars) (Figure 5.3). The rebars are vertical, horizontal, and diagonal. Continuity of the rebars at splices may be provided by metal-filled butt splices that provide a joint as strong as the bars. The foundation mat reinforcement consists of top and bottom layers of rebars.

How strong are the openings in the concrete, the penetrations?

Nuclear reactor containments have large penetrations such as the access air locks and the equipment hatches. The reinforcing steel around these

Figure 5.2 Containment steel liner with studs.

Figure 5.3 Containment rebars.

openings is typically bent, with supplementary bars added to sustain stress concentrations and redistribute stress at these openings. There are smaller penetrations for the fuel transfer tube, piping (Figure 5.4), electrical cables, and instruments.

What are the different types of embedments?

There are four general types of embedded steel plates:

- Strip plates that are in the range of 3/4″ thick, 8″ or 10″ wide with various lengths. They are mounted with studs about 3/4″ diameter

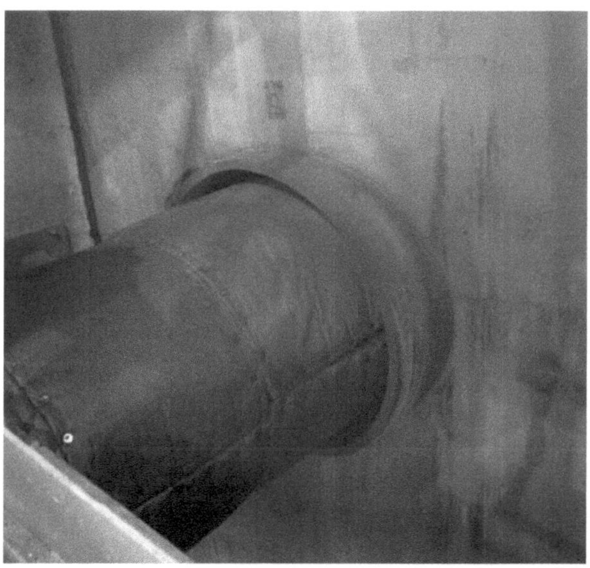

Figure 5.4 Wall penetration for an insulated pipe.

into concrete walls, columns, beams, and slabs. The studs are welded to the strip plate typically in pairs 1 ft apart.

- Sheet plates that are in the range of 3/4″ thick, with varying lengths and widths mounted to concrete walls. The plates are anchored to the concrete with studs in manner similar to the strips.
- Dedicated embedments are another form of steel plates that are neither strip plates nor sheet plates.
- Angles are steel shapes flush-mounted in concrete walls and floors, and to which equipment can be directly welded.

How are these structures qualified?

The containment structure is analyzed and qualified by large finite element models that include the soil layers, the main internal walls of the structures, and a simplified model of the reactor coolant loop and primary equipment.

Scale tests of reinforced containments were conducted in Japan on a large shake table to verify the numerical predictions. From a structural strength point of view, these structures have proved to resist well the shaking from actual large earthquakes, in some cases well above the SSE.

How is a concrete containment analyzed for a postulated pipe break inside containment?

The assumed break of a large pipe inside containment, for example, a main steam line, causes a spike in pressure and temperature inside the containment. The challenge is to first accommodate the immediate internal spike in pressure from the escaping steam, which causes the liner to bear against the concrete. Then, it is also necessary to allow for the differential expansion between the steel liner and the concrete walls whose temperature lags behind. It becomes a design engineering challenge of strain compatibility between the liner and the reinforcing steel. The concrete may crack under these conditions but must be capable of sustaining the break loads and not impair the pressure retention capability of the steel liner.

How is the liner attached to the concrete?

The liner is anchored to the steel reinforcement in the concrete, and it is thicker near stress concentrations such as penetrations and near load concentrations such as the large containment polar crane.

Is the leak tightness of the liner checked periodically?

Yes, this is a requirement of the in-service inspection code ASME XI. The containment liner and penetrations are designed and constructed to permit periodic integrated leak rate tests during the plant lifetime. Fatigue analysis of these repeated tests assumes a conservatively large number of pressure cycles.

What are other design loads for the liner?

There are several loads to include in designing the liner: its weight; the deformation of the concrete over time; possible ground settlement; seasonal temperature changes; hydrostatic pressure from flooding; differential expansion from all sources, including a postulated pipe break and the resulting overpressure from escaping steam; the loads from earthquake shaking cycles, tornado winds, and missiles; large impacts; and in some cases explosions. In some plants, the postulated high-energy line breaks result in large direct whip impacts and jet impingement loads on the liner.

Other than on the containment, where are liners used?

The spent fuel pool and the refueling cavity have steel liners to prevent leaks from seeping through concrete walls and to protect the concrete against radiation damage from the decaying spent fuel. Steel liners are also used to protect concrete tanks such as the refueling water storage tank, the condensate storage tank, and the reactor water storage tank. In these cases, the primary function of the tank liner is to provide a watertight barrier.

What are the other main concrete buildings?

Beyond the reactor building and its internal compartments and sub-compartments, there are several other concrete buildings in the plant: the fuel building that houses the spent fuel and the building that houses the emergency core cooling pumps, the containment spray pumps and coolers, the emergency feed pumps, the service water pumps, the emergency diesel generators, the electrical switchgear, and the emergency generators.

5.3 STRUCTURAL WELDING

When is welding of structural steel preferred to bolting?

There are basically two methods for joining steel members to support equipment or distribution systems: welding (Figures 5.5–5.7) or bolting (Figures 5.8–5.12). Following are considerations for choosing one method over the other:

- From an economical and versatility standpoint, welded connections are generally more readily implemented in the field. For example, welding

Figure 5.5 Examples of fillet welded connections of tube steel to baseplate.

Figure 5.6 Pipe hanger member welded to steel beam.

two tube steel sections together at 90° is more feasible than trying to bolt them together.

- Welding of steel members is selected where bolting would not be practical due to interferences, where higher strength and rigidity of the joint are needed.
- Welding would be preferred when the design is permanent and does not need to be disassembled. Example of welded joints include a welded tube steel to a baseplate (Figure 5.5), a welded pipe support member to a baseplate (Figure 5.6), or a welded steel frame pipe support (Figure 5.7).

Figure 5.7 Flare bevel welds between tube sections.

Figure 5.8 Side view of bolted structural steel.

Figure 5.9 Edge view of bolted structural steel.

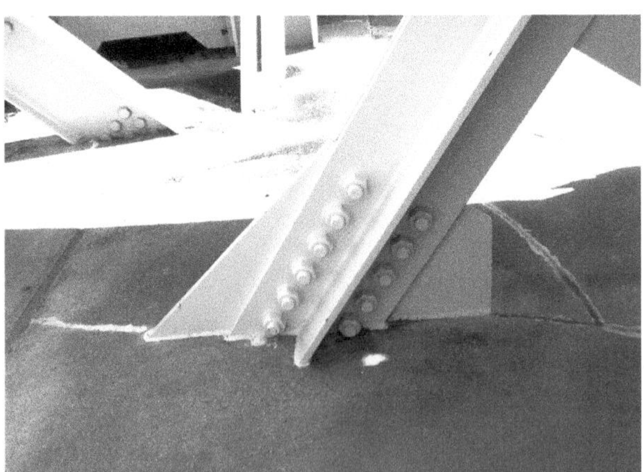

Figure 5.10 Bolted structural steel joint embedded in concrete.

Figure 5.11 Bolted structural steel bracing.

Figure 5.12 Structural steel beams and columns.

- Bolted structural joints can be designed to allow a degree of freedom in rotation like a pin joint, where some flexibility may be desired, while welding behaves like a fixed end and minimizes deflections and rotations.

What weld electrodes are most commonly used in welding structural steel in a nuclear power plant?

First, let us take a moment to examine the base metal for structural steel. The common choices for carbon steel structures are ASTM A572 for high-strength applications, ASTM A36 for general applications, ASTM A500 for structural tubing (tube steel), and ASTM A501 for structural pipe. The common choice for stainless steel is ASTM A240 type 304L or similar.

Welding electrodes are typically E70XX for welding carbon steel and E60XX in the older plants. For welding stainless steel, a common weld electrode is E308L-16.

How does the strength of a weld compare to that of the base metal?

The strength of the weld filler metal normally overmatches that of the base metal. This does not necessarily imply that the weld is stronger than the base metal, since the weld strength depends on its shape, size, and quality. The quality of the weld depends, in turn, on the weld method, the weld procedure, the welder, the heat treatment (where applicable), and the weld examination and testing. In summary, the strength of a weld and (as importantly) its toughness depend on its quality, whereas the strength and toughness of base metal do not involve these field construction variables.

What codes govern the design of structural welds?

For safety-related support structures, the weld design is governed by ASME III Division 1 Subsection NF. For safety-related, seismic category I or II/I steel structures AISC N-690 may apply, in some cases, with supplementary project-specific requirements.

For nonsafety-related support structures, AISC-S335 "Buildings, Structural Steel, Allowable Stress Design and Plastic Design" and AWS D1.1 "Structural Welding Code for Acceptance Criteria" generally apply, with AWS A2.4 for welding symbols and AWS D9.1 for welding of sheet metal. Weld sizes and effective areas for weld qualification and strength can be determined following the procedures in AISC.

Nuclear Construction Issues Group document NCIG-01 "Visual Weld Acceptance Criteria for Structural Welding at Nuclear Power Plants" has also been used as a reference for welding examination.

What if a steel support member is welded to an ASME III pressure boundary such as a vessel, a valve, a pump, or a pipe?

This is the case, for example, for pipe anchors, rectangular lugs, or cylindrical trunnions. In this case, the weld has to meet the design and fabrication requirements of the pressure boundary component. In particular, the weld has to be designed, sized, and analyzed for all the loads transmitted by the equipment. The rules for the analysis of welded attachments to ASME III pressure boundaries were first published in WRC Bulletins, which then became Code Cases, and are now included into the body of ASME III as Appendix Y.

Are there advantages in using one weld configuration over another?

A full penetration weld provides the greatest strength, as it uses the full cross-section of the base metal and avoids the stress concentrations and crevices of partial penetration welds. But in most cases steel members are welded together by fillet welds, which are sized to the requirements of ASME III NF or AISC.

Other forms of welding are used in specific applications: For example, plug or slot welds are used to add additional strength to an existing weld. Repair weld deposition is used to fill defects, to round discontinuities, or to restore base metal. Tack welding, which is a temporary weld, is normally applied for alignment purposes before completing the weld.

Are there special precautions regarding welding of safety-related structures in nuclear power plants?

First, it is important to understand the jurisdictional boundaries between steel structures and ASME III NF equipment supports, since in many cases steel structures are erected to support equipment. Understanding these boundaries is facilitated by a series of figures in ASME III NF-1130.

If the steel structure is a support to an ASME III component (pipe, valve, pump, vessel), then the support structure is welded in accordance with ASME III Subsection NF, in particular, NF-4240 "Requirements for Welded Joints," NF-4300, and NF-4400. The examination would be in accordance with NF-5000.

These safety-related welds are made in accordance with an ASME IX qualification program by an ASME certificate holder. In practice, in safety-related applications, every structural weld is defined and sized in the engineering design and documented in fabrication and erection drawings. NF-4000 addresses the welding practice, and the welding program and procedures.

What are the requirements, or practices, for welding nonsafety-related structures?

Welding of nonsafety-related structures in a nuclear power plant is similar to welding in other critical industries. Welding typically follows AWS D1.1. Written programs address the design of the weld joint, the selection of the welding technique and weld material, the qualification of the procedure and personnel, the quality of implementation, and the examination of the welds. These activities are given particularly close attention in a nuclear power plant, even for nonsafety joints.

How are welds sized?

For welded steel supports in safety-related applications, NF-3133 requires that a design report be prepared for support structures to show compliance with stress limits in NF-3000. For nonsafety-related structures, welding design and qualification can be performed in accordance with the AISC Manual.

For NF support structures, there are requirements for Class 1 welds (NF-3226) and different requirements for Class 2 and 3 (NF-3256). These subsections spell out the allowed types of weld joints and the allowable stresses.

Welds are sized by first calculating the loads (forces and moments) at the welded joint. This can be straightforward for a simple structure, but elaborate for a complex frame, which will necessitate an FEA. The loads are then converted into stresses, following the requirements of the applicable code (ASME III NF or AISC). Alternatively, the weld may be treated as a line (the classic approach described in "Design of Welded Structures" by Blodgett), and the applied force per unit length is calculated and compared with an allowable based on the lowest strength of the weld and base metal material; this procedure results in the weld size needed to resist the load.

In addition to the calculated and qualified weld size, the codes and engineering practice often impose a minimum and maximum weld size. Actual weld sizes are usually in multiples of 1/16″. The weld size is determined by the thicknesses of the joint pieces. Typically, a weld size would not exceed the thickness of the thinner part welded. The weld size and its length are then determined to achieve the required strength of the welded joint.

5.4 BOLTING TO STEEL

What types of bolts are used in component support structures?

There are two general types of bolts used in power plant structures: (1) bolts used to join structural steel members or (2) concrete anchor bolts used to join steel baseplates or structural shapes to concrete.

What type of bolts are most often used to join structural steel members?

Two categories of bolts are available to join structural steel. They include (1) regular-strength bolts in snug-tight applications or (2) high-strength bolts:

- A regular-strength bolt, such as ASTM A307, with lock washers, is usually used where a snug-tight condition is sufficient to join lightweight

structures. A snug-tight joint is achieved by using an ordinary wrench or an impact wrench, followed by a tightness check.
- A high-strength bolt, such as ASTM A490 or ASTM A325, is used for joints that have to sustain large loads. These bolts rely on prespecified preload and exacting installation to achieve the preload, either by a specified turn of the nut or by using a torque wrench. In the turn of the nut method, the nut is turned to a specified angle from the tight condition. If the preload is achieved by an applied torque, the preload will be a function of the applied torque, the friction factor, and the diameter of the bolt. The preload is based on the strength of the material in tension and is limited to a specified allowable preload tensile stress as a function of the material yield strength.

The strength of bolted joints relies on friction or bearing. A friction joint depends on a sufficiently high clamping force to prevent slippage of the connected parts under the applied loads. In this case, the relative movement between joints is not permitted. A bearing joint depends on the contact of the fasteners against the side of their holes to transfer the load between the connected parts. Bolting material in this application can come in two forms: a high-strength bolt (with a hardened washer and nut) or a screw.

What codes govern structural steel bolting?

For nuclear power plant applications, the commonly used codes are AISC "Specification for the Design, Fabrication, and Erection of Structural Steel for Buildings," AISC N-690 "Specification for Safety-Related Steel Structures for Nuclear Facilities," "Specification for Structural Joints Using ASTM A325 or A490 Bolts" by the Research Council on Riveted and Bolted Structural Joints (RCRBSJ), and ASME III, Subsection NF.

The codes for structural steel bolting govern preload, load combinations, and allowable stresses in tension and shear. The codes also provide important considerations for structural bolting to prevent failure of the connecting parts: preload, minimum edge distance, and minimum spacing between fasteners. ASME B1.1 and B18.2.1 govern bolt details related to thread form, series, class, allowance, tolerance, and designation for unified screw threads, dimensional requirements, and grade markings for bolts and screws.

5.5 BOLTING TO CONCRETE

How many types of concrete anchor bolts are there?

There are many. But any classification should start by differentiating the two main categories of concrete anchor bolts: the cast-in-place bolts and the postinstalled bolts. Cast-in-place bolts are positioned in place before the concrete is poured, while postinstalled anchor bolts are drilled in position after the concrete is cured. Beyond these two main categories, concrete anchor bolts come in a variety of forms and functions: special inserts, drilled-in expansion anchors, embedded anchor bolts, through bolts, grouted anchors, liner anchors, penetration anchors, and embedded corner angles. Each type has its purpose and application, and manufacturer requirements for selection, design, qualification, and installation.

What is the function of concrete anchor bolts?

The function of a concrete anchor bolt is to anchor down attachments to concrete surfaces or blocks and to resist the applied loads. The loads can be either service loads (such as weight and thermal expansion) or postulated loads (such as seismic and pipe rupture). These loads are resisted by the tension and shear capabilities of the bolted joint.

What codes govern concrete anchor bolts?

The common codes for the design, analysis, and qualification of concrete anchor bolts are ACI-318 "Building Code Requirements for Reinforced Concrete" for nonsafety-related bolts and ACI-349 "Code Requirement for Nuclear Safety Related Concrete Structures" for safety-related bolts, as well as bolt manufacturer guides and publications. Because the ACI formulas and methods for the design of concrete anchor bolts have changed over the years, it is essential in nuclear plant applications that the design, analysis, and qualification of concrete anchor bolts follow the licensing basis of the plant, including the applicable edition of the ACI code.

What is the design margin for concrete anchor bolts?

The design margins on expansion anchors were addressed in NRC Bulletin 79-02 "Pipe Support Base Plate Designs Using Concrete Expansion Anchors." The minimum design margin is 4 for wedge and sleeve type anchor bolts, and a margin of 5 is imposed for shell anchor bolts.

Why is the bolt itself designed to fail before the concrete?

The failure of concrete anchor bolts involves either the failure of the concrete itself or the failure of the bolt itself. The overload of concrete will result in a sudden brittle fracture of the concrete that does not allow for load redistribution; while the overload of the steel bolt itself will cause it to elongate plastically that will cause a redistribution of the applied load to the other less-loaded bolts, this will delay failure or may prevent it altogether. Therefore, the bolt design, material, embedment depth, spacing from one another and from free edges should be such that the bolt itself rather than the concrete governs the design, bringing into play the steel's plastic elongation that would permit load redistribution and possibly preventing failure.

What parameters govern the concrete anchor bolt capacity?

The ultimate capacity of a concrete anchor bolt is governed by the quality of installation, and the bolt material, its size, its embedment depth, the anchor-to-anchor distance, the anchor-to-concrete edge distance, the quality and strength of the concrete, the presence of cracks in concrete, and the possible moment arm caused by an offset between the baseplate and the concrete. The use of thin, flexible baseplates (less than 1″ or $1^1/_2$ ″ thick) can lead to an additional bolt tension as a result of plate bending, an effect known as prying action.

How to set the tensile capacity of closely spaced bolts?

The tensile capacity of concrete is based on a uniform tensile stress acting on an effective stress area; this is the projected area of a stress cone radiating from the tip of the anchor toward the concrete surface. The cone approach has been replaced by a pyramidal shape in the newer editions of ACI. The effective area of the concrete capacity is limited by overlapping concrete stress cones, by the intersection of cones with adjacent concrete edges, and by the overall thickness of the concrete. For closely spaced anchors of the same size and type, where stress cones overlap, a reduction in concrete capacity must be considered.

When anchors are closely spaced such that they cause a reduction in capacity, the reduced capacities may be calculated by considering a bolt group failure. The outer radius of the bolt group is set equal to the depth of embedment. To prevent exceeding the concrete capacity, the total load on the bolt group is reduced to the design capacity of the total effective concrete stress area. At any point, the capacity of each anchor should not exceed the capacity of the isolated anchor.

How does edge distance affect the bolt capacity?

Anchors that are near a free edge of the concrete and loaded in tension or shear can cause a spalling failure of the edge, as has been observed in practice and tests. Figure 5.13 illustrates the earthquake-induced failure of a concrete pad supporting a tank (in this case a winery tank), which was

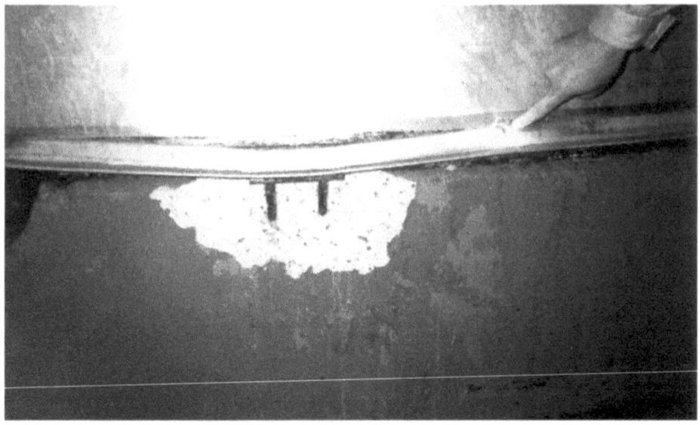

Figure 5.13 Earthquake-induced concrete failure from insufficient edge distance.

caused by insufficient distance between the embedded steel and the concrete edge.

The use of a minimum distance from the edge is therefore necessary to confine the lateral load transfer from the steel to the concrete. In the case of expansion anchors located closer than one embedment depth from a free edge or penetration, the reduced effective concrete stress area may be calculated using methods based on overlapping cones, by considering, for example, a linear reduction based on the ratio of the free edge distance to the depth of embedment. The relationship between embedment depth and edge distance can also be replaced by a number of bolt diameters to the edge.

For anchor bolt connections where the failure is based on the tensile capacity of the steel, the allowable tension force is a function of the yield or tensile strength of the bolt material multiplied and the cross-sectional tensile area of the anchor.

How to use steel plates embedded in concrete?

Embedded plates are steel plates, normally less than 1″ thick and less than 12″ wide, with varying lengths, mounted in concrete walls, columns, beams, and slabs, using studs welded to the plate (Figures 5.14 and 5.15).

The spacing between the studs, the edge distance, and the stresses in the concrete are critical considerations for the design of embedded plates to achieve the desired stiffness.

An issue that arose in the 1970s had to do with the importance of the stiffness of the embedded plates under load and the potential for prying. This issue has been addressed in the nuclear industry since the publication of NRC Bulletin 79-02 "Pipe Support Base Plate Designs Using Concrete Expansion Anchors."

Figure 5.14 Embedded steel plates.

Figure 5.15 Strip plates with studs.

5.6 CASE STUDIES

Oftentimes the plant engineer is tasked with finding a cost-effective and safely conservative solution to emerging structural and mechanical questions: making a modification, using an existing structural member to lift a piece of equipment, modifying loads on a structure as a result of operating or maintenance modifications. Today, many of these structural questions can be solved by detailed finite element structural analyses. But, in the short time frames that are typical of a plant engineer's everyday life, it is essential to be able to address these requests using the fundamental principles of statics and dynamics, coupled with the margins in plant-specific codes, standards, and regulatory requirements. Our objective in these case studies is to illustrate some of these quick turnaround assignments.

Case study 1: Equipment Lift

The plant needs to lift a 2.5 kips manway in the turbine building to provide access for inspection. The plant has an ASTM A36 W8 × 31 I-beam of the right length, which can span three existing support points (Figure 5.16).

The beam parameters are as follows: span length, $L = 64$ in; linear weight of I-beam, $w = 31$ lb/ft; lift load including a safety factor of 2 on the load to be lifted, $P = 5$ kips; the weight that is lifted at midspan, $a = b = L/2 = 32$ in; elastic section modulus of W8 × 31, $S_x = 27.5\ \text{in}^3$; cross-sectional area of W8 × 31, $A = 9.13\ \text{in}^2$; and minimum specified yield strength of the beam at ambient temperature, $F_y = 36$ ksi. The adequacy of the beam will be checked using the AISC working stress method (i.e., elastic design method).

The maximum bending moments in the beam due to P and due to w are

$$M_{max,P} = \frac{13}{64} PL = 65 \text{ in kips } \textit{at point } P; \textit{ and}$$

$$M_{max,w} = \frac{wL^2}{8} = 1.3 \text{ in kips } \textit{at point } R_2$$

The corresponding bending stress, conservatively adding the maximum moments, is

$$f_b = \frac{M_{max}}{S_x} \cong 2.4 \text{ ksi} < 0.6\ F_y(\textit{AISC allowable bending stress}) = 21.6 \text{ ksi}$$

The maximum shear is

$$V_{max,P} = \frac{19}{32} P = 3.0 \text{ kips } \textit{at point } P; \textit{ and}$$

$$V_{max,w} = \frac{5wL}{8} = 0.1 \text{ kips } \textit{at point } R_2$$

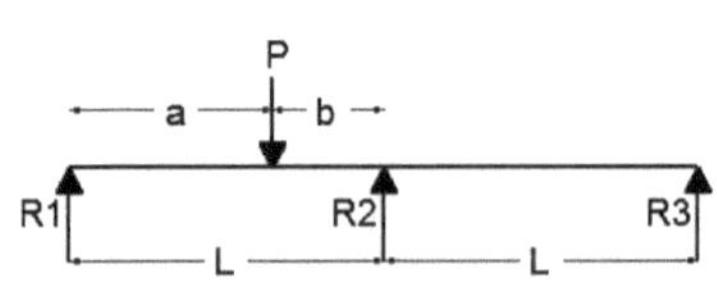

Figure 5.16 Lift beam diagram.

The corresponding shear stress, conservatively adding the maximum shear forces, is

$$f_v = \frac{V_{max}}{A} \cong 0.3 \text{ ksi} < 0.4\ F_y\ (\textit{AISC allowable shear stress}) = 14.4 \text{ ksi}$$

The deflection of the beam is calculated to be small, using either AISC Part 2 "Camber and Deflection Coefficients" or the exact beam equation solution.

Case study 2: Tornado Pressure on Hatch Cover

The vertically mounted equipment hatch cover has been designed to act as a tornado barrier to resist the differential pressure from the postulated tornado, the tornado missile, and, separately, a seismic SSE load. The hatch cover is secured to the wall by 32 ASTM A325 bolts $1^1/_2$ in in diameter (bolt nominal diameter $A_{bolt} = 1.767$ in^2 per AISC Part 4). The hatch opening is 16 ft in diameter, and the cover is $D = 19$ ft in diameter (Figure 5.17). The hatch cover is 18-in thick and consists of 16 in of reinforced concrete with a 2-in steel plate and weighs $W = 80$ kips. The plant would like to have the option of untorquing half the bolts (16 bolts) to expedite outage operations.

The tornado builds up a differential pressure across the hatch cover equal to $\Delta P = 3$ psi, according to the plant FSAR Chapter 2. The dynamic load factor is 1.0 in this case because the rise time of the tornado differential pressure is much larger than the natural period of the hatch in the out-of-plane direction. The tornado-induced tension force in each of the remaining 16 bolts is limited to the AISC Part 4 load limit for $1^1/_2$ A325 bolts:

$$F_{t,tornado} = \frac{\pi D^2}{4} \frac{\Delta P}{16\ bolts} = 7.7 \text{ kips} < 77.7 \text{ kips} \rightarrow \textit{acceptable}$$

The local effects of a tornado missile (penetration, scabbing on the back face, spalling on the impacted face) are not affected by the 16 bolts. The global effects of a tornado missile (bending of the cover from direct impact) cause a reaction tension in the bolts at the periphery, which should be addressed. The resultant radial SSE acceleration (placing the bolts in tension) is calculated to be $a_R = 1$ g, and the resultant vertical SSE acceleration (placing the bolts in shear), including 1 g gravity, is calculated to be $a_V = 2$ g. The SSE stress in each of the remaining 16 bolts is limited

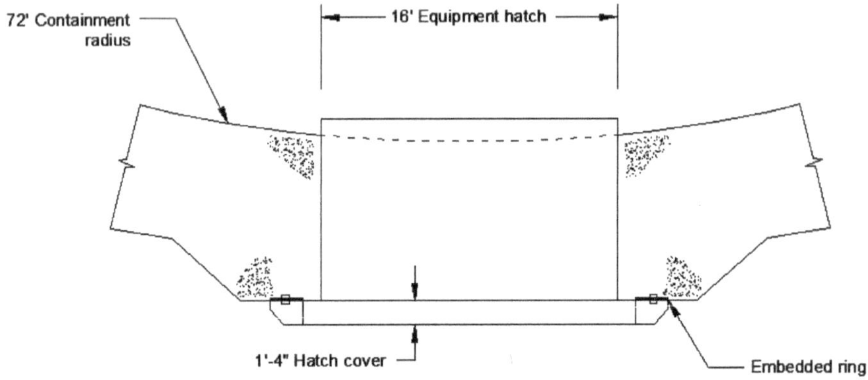

Figure 5.17 Schematic of hatch cover.

by the AISC Part 4 stress limit for A325 bolts of 44.0 ksi in tension and 17.0 ksi in shear:

$$\frac{(a_R W)}{16\ bolts \times A_{bolt}} = 2.8\ \text{ksi} < 44.0\ \text{ksi} \rightarrow \textit{acceptable}$$

$$\frac{(a_V W)}{16\ bolts \times A_{bolt}} = 5.7\ \text{ksi} < 17.0\ \text{ksi} \rightarrow \textit{acceptable}$$

The tension–shear interaction is also checked to be acceptable with 16 bolts. In this case, the untorquing of half the bolts, leaving them in position, can be implemented as the plant approaches the refueling shutdown (Figure 5.17).

Case study 3: Welded Steel Support Structure

An existing welded steel structure was designed for a pipe reaction load $F_y = 250$ lb (Figure 5.18(a)). The existing valve near the support was replaced by a new, heavier valve, which brought the total load on the support up to $F = 424$ lb. Engineering is asked whether the existing support can take the new increased load. In this example, there is no lateral pipe load on the support, only the vertical downward load of 424 lb.

The demand vs capacity check must address each member in the load path: the U-bolt, the horizontal L-shaped 3 × 3 × 3/8 angle, the weld between the L and the 4 × 4 × 1/4 tube steel, the vertical tube steel, the weld between the vertical tube steel and the baseplate (Figure 5.18(b)), the baseplate itself, and the concrete anchor bolts.

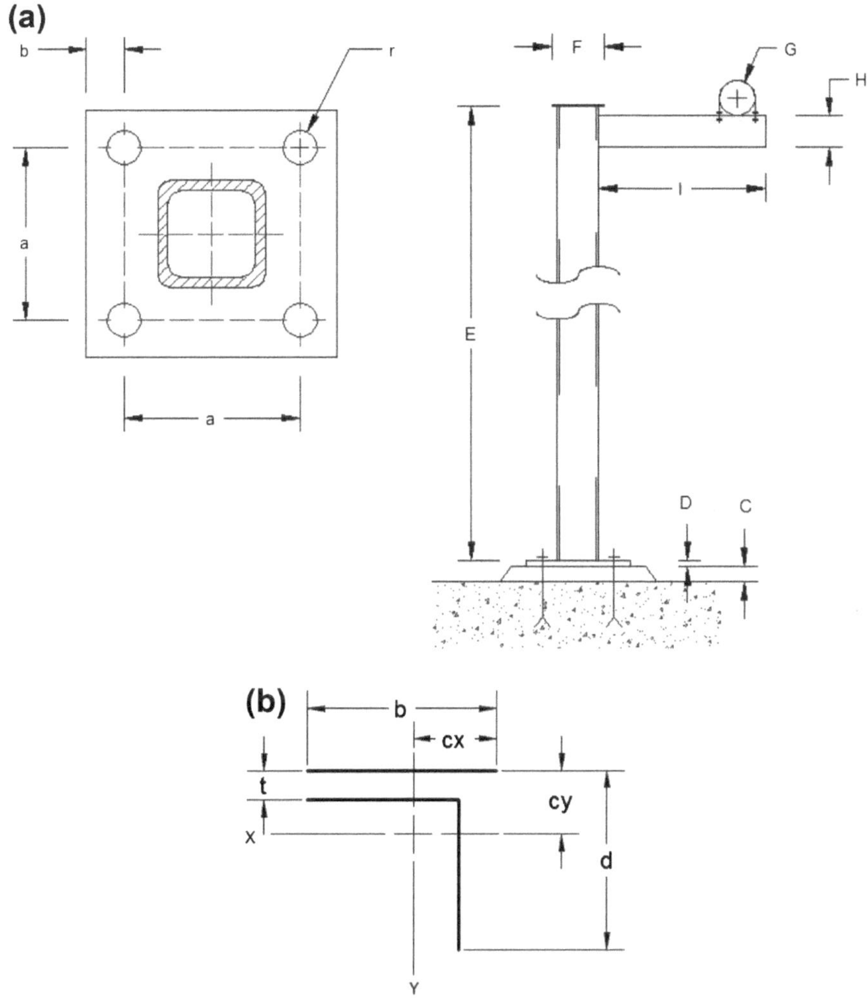

Figure 5.18 (a) Pre-walkdown schematic of the support showing dimensions to collect. (b) Weld dimensions.

This is a standard series of calculations, following AISC "Manual of Steel Construction" and ACI-318 "Building Code Requirements for Structural Concrete." The weld is analyzed as a line, following the guidance of the classic textbook by Omer W. Blodgett "Design of Welded Structures."

The U-bolt

In this case, the U-bolt is $^1/_2$ in in diameter, from manufacturer ABC. It is acceptable as is, since it is simply a guide and does not have to react a lateral load. When a U-bolt must react lateral loads, the load capacity must be obtained from the bolt manufacturer.

The Horizontal L *3 × 3 × 3/8 Angle*

The maximum bending stress is

$$f_{bL} = \frac{F \times l}{S_L} < 0.6\ F_y$$

Substituting $F = 424$ lb, $l = 17$ in (this is the design eccentricity of 15″ plus a tolerance of 2″ on the point of application of the load), $S_L = 0.833\ \text{in}^3$ section modulus of the $L\ 3 \times 3 \times 3/8$, and $F_y = 36$ ksi for ASTM A36 steel, we obtain $f_{bL} = 8.7 < 21.6$ ksi → qualifies. Note: In this case, the weight of the $L\ 3 \times 3 \times 3/8$ angle is 7.2 lb/ft, which is small compared to the applied load $F = 424$ lb.

The maximum shear stress is

$$f_{vL} = 2\frac{F}{A_L} < 0.4\ F_y$$

Substituting $F = 424$ lb, $A_L = 2.11\ \text{in}^2$ cross-sectional area, and $F_y = 36$ ksi, we obtain $f_{vL} = 0.4 < 14.4$ ksi → qualifies.

The deflection of the L angle is

$$\Delta_{max} = \frac{F \times l^3}{3 \times E \times I_L} < \frac{1}{16}\text{in}$$

Substituting $F = 424$ lb, $l = 17$ in, $E = 29{,}000$ ksi, $I_L = 1.76\ \text{in}^4$ moment of inertia, we obtain $\Delta_{max} = 0.01 < 1/16$ in → qualifies.

The Weld between L *3 × 3 × 3/8 and* TS *4 × 4 × 1/4*

The horizontal L angle is welded to the vertical tube steel with a three-sided $^1/_4$-in weld, as indicated in the figure. Following O. Blodgett, treating the weld as a line, we calculate the weld properties.

$$a = b - t; \quad e = d - t$$

$$L_w = 2a + d$$

$$c_x = \frac{b^2 + t(d - 1.5t)}{L_w}; \quad c_y = \frac{t(2a - t) + d^2}{2L_w}$$

$$I_x = bc_y^2 + a\left(c_y - t\right)^2 + \frac{e^3}{12} + e\left(\frac{d+t}{2} - c_y\right)^2$$

$$I_y = \frac{a^3 + b^3}{12} + b\left(\frac{b}{2} - c_x\right)^2 + a\left(\frac{b+1}{2} - c_x\right)^2 + e(c_x - t)^2$$

$$J_w = I_x + I_y$$

Substituting $b = 3$ in, $d = 3$ in, $t = 3/8$ in, we obtain $L_w = 8.25$ in, $I_x = 5.8$ in^4, $I_y = 7.5$ in^3, and $J_w = 13.3$ in^3.

Having calculated the weld properties, we calculate the loads on the weld group

$$F = 424 \text{ lb}; \quad M_x = F \times l = 7.2 \text{ in kips}; \quad M_z = F(c_x - t) = 0.4 \text{ in kips}$$

We can now calculate the maximum stresses in the welds

$$f_x = \frac{M_z\left(d - c_y\right)}{J_w} = 0.06 \text{ kip/in}$$

$$f_y = \frac{F_y}{L_w} + \frac{M_z(b - c_x)}{J_w} = 0.10 \text{ kip/in}$$

$$f_z = \frac{M_x\left(d - c_y\right)}{I_x} = 2.91 \text{ kip/in}$$

$$f_w = \sqrt{f_x^2 + f_y^2 + f_z^2} = 2.92 \text{ kip/in}$$

The allowable weld stress is

$$F_w = \min\left\{0.3 \times 0.707 \times F_u \text{ ksi} \times 0.25 \text{ in}; 0.4 \times F_y \text{ ksi} \times 0.25 \text{ in}\right\}$$
$$= 3.60 \text{ kip/in}$$

The demand over capacity (maximum calculated stress over allowable stress) is therefore

$$\frac{f_w}{F_w} = 0.81 < 1 \rightarrow \textit{qualifies}$$

The Vertical 4 × 4 × 1/4 Tube Steel

The vertical tube steel is made from ASTM A500 "Standard Specification for Cold-Formed Welded and Seamless Carbon Steel Structural Tubing in Rounds and Shapes" with a yield stress $F_{y\text{-}A500} = 46$ ksi. The axial stress f_a and bending stress f_b in the tube are

$$f_a = \frac{F + W}{A_{TS}}$$

$$f_b = \frac{M_x + M_z}{S_{TS}}$$

Substituting $F = 424$ lb, the applied load; $W = 0.11$ kip, the self-weight of the 9-ft-long tube steel; the moments $M_x = 7.2$ in kip and $M_z = 0.4$ in kip have been calculated above; $A_{TS} = 3.59$ in^2, the cross-sectional area of the tube steel; $S_{TS} = 4.11$ in^3, the section modulus of the tube steel, we obtain $f_a = 0.2$ ksi and $f_b = 1.9$ ksi. Normally these stresses would be compared with the AISC allowable stresses and the buckling limit. For an engineer who has performed these calculations many times, it is evident that these applied stresses are clearly within the AISC allowable limits for vertical members in compression and bending.

There is however one stress in the tube steel that needs to be checked, it is the local stress in the tube steel wall at the weld to the horizontal angle. The stress from the horizontal leg is

$$f_{b-weld} = \frac{M_x c_y}{I_x}$$

Substituting $M_x = 7.2$ in kips, applied moment; $c_y = 0.66$ in, weld size; and $I_x = 6.44$ in^3, moment of inertia; we obtain $f_{b\text{-}weld} = 0.7$ ksi. We now calculate the local stress in the tube steel wall:

$$M_L = \frac{f_{b-weld} \times (3.5 \text{ in})^2}{12} = 0.8 \text{ in kip}$$

$$S = \frac{(12 \times 0.25 \text{ in})(0.25 \text{ in})^2}{6} = 0.03 \text{ in}^3$$

$$f_b = \frac{M_L}{S} = 26.6 \text{ ksi} < 0.75\, F_{y-A500} = 34.5 \text{ ksi} \rightarrow \textit{qualifies}$$

The Weld between the Tube Steel and the Baseplate

The weld between the 4 × 4 × 1/4 vertical tube steel and the ½ -in baseplate is a 3/16-in all-around fillet weld. This weld is subjected to a bending stress given by

$$f_{weld} = \frac{F + W}{L_w} + \frac{M_x + M_z}{S_w}$$

Substituting $F = 424$ lb, the applied load; $W = 0.11$ kips, the weight of the tube steel; $L_w = 15$ in, the weld length with margin; $M_x = 7.2$ in kips and $M_z = 0.4$ in kips, the applied moments; and $S_w = 19.5$ in^2, the section modulus of the weld; we obtain $f_{weld} = 0.5$ kip/in, which is quite low and well within the AISC limit for the 3/16-in fillet weld.

The Concrete Anchor Bolts

The baseplate is anchored to the concrete by four 5/8-in-diameter anchor bolts from manufacturer XYZ. There is no shear on these bolts, only tension caused by overturning, equal to

$$T_{max} = \frac{M_x + M_z}{s}$$

$$s = \frac{a}{2} + \frac{4\text{ in}}{2} + 2 \times 0.25\text{ in}$$

Where $a = 7$ in is the distance between bolts and 0.25 in is the plate thickness, we added half the tube steel as 4 in/2 to reflect that the moment is calculated around an edge of the tube steel. The moments M_x and M_z are calculated above. Substituting, we obtain the tension in a bolt as $T_{max} = 0.63$ kips, compared to the catalog limit of 2.75 kips → qualifies. Note that for anchor bolts we would have to check a few characteristics to make sure that the full catalog capacity of 2.75 kips applies: (1) the embedment length of the anchor bolt in the concrete, (2) the tightness of the bolt, since it is an existing bolt, (3) the anchor bolt distance to the edge of the concrete, (4) the distance from bolt to bolt, and (5) the lack of cracks in the concrete.

Note: Tests conducted in 1986 on installed anchors showed that their actual capacity was less than the catalog values. This was the subject of Information Notice 86-94 and was resolved by further tests and revision to the catalog capacities.

If there was a shear load on these anchor bolts, we would have to check the shear load by itself, and the interaction of tension and shear on the bolt.

Finally, because the baseplate is grouted, the shear load will cause bending in the bolt that must be considered.

The Baseplate

The baseplate is made of ASTM A36 and is $10 \times 10 \times {}^1/_2$ in thick. The bending stress in the plate $f_{b-plate}$ is given by

$$f_{b-plate} = \frac{1.5 \times M_{plate}}{S_{plate}}$$

Where

$$M_{plate} = 2\ T_{max} \frac{b_{plate} - 1.5 \text{ in}}{2}$$

$$S_{plate} = \frac{b_{plate} \times t_{plate}^2}{6}$$

Substituting $t_{plate} = 0.5$ in and $b_{plate} = 10$ in, we obtain $S_{plate} = 0.42$ in^3, $M_{plate} = 5.4$ in kips, and $f_{b-plate} = 19.3$ ksi $< 0.75\ F_{y-A36} = 27$ ksi $\rightarrow$ qualifies.

In Summary

The members in the load path qualify: the U-bolt, the *L* angle, the *L* angle to tube steel weld, the tube steel, the tube steel at the bolt, the concrete anchor bolts, and the baseplate.

Case Study 4: Pipe Whip into a Structural Beam

A 10-in Scheme 160 SA-312 Type 316 stainless-steel piping system operates at 2000 psi and 600 °F. The pipe is rerouted, and as a result of the reanalysis we have to postulate a new circumferential break location at an elbow.

The new break causes a pipe whip and impact with a structural beam a distance a = 5 ft away from the pipe Figure 5.19. We need to calculate the impact energy E_{impact} transmitted by the whipping pipe to the target beam, and from there the dynamic force F_{beam} on the beam and its attachments to the building structure. These loads will be used to check the adequacy of the existing beam and its backup structure.

$$F_{thrust} = RF \times TF \times P \times A$$

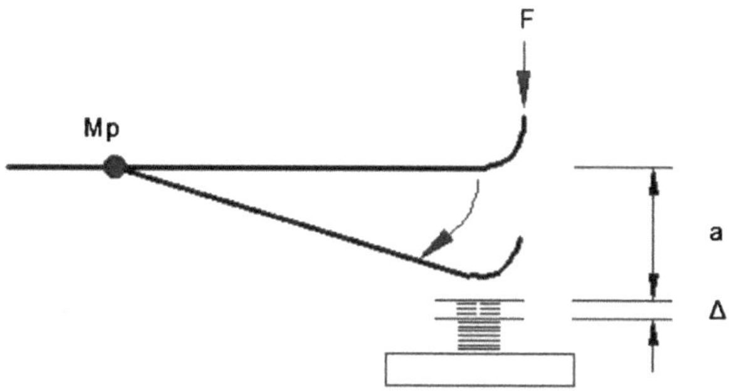

Figure 5.19 Pipe whip into structural beam.

Where F_{thrust} is the hydraulic thrust force from the discharging liquid; RF is a dynamic rebound factor of 1.1; $TF = 1.26$ is a thrust factor for flashing steam from ANS 58.2; $P = 2000$ psi is the initial pressure; $A = 56.7$ in^2 is the pipe flow area; In this case F_{thrust} = 157 kips, which is a very large load. This load will typically cause a plastic hinge in the pipe. The pipe will then whip around the hinge and impact the first intervening structure in its path.

If the intervening structure or the backup structures do not qualify for this pipe whip impact force, there are several options for resolution: (1) Preventing the impact by supporting the line to reduce its pipe break postulation stresses and therefore not to have to postulate the intermediate break at the elbow to start with. (2) Placing a tight gap whip restraint near the elbow to prevent the whip and therefore the impact. This however is only possible if the new tight gap whip restraint does not impede thermal expansion of the line. (3) Installing an energy-absorbing whip restraint with a sufficient gap, if congestion and layout permit. (4) Placing a crush-pipe bolted to the target structure to plastically absorb the brunt of the impact. (5) Rerouting the target, if at all feasible. (6) Analyzing the target using plastic analysis or ductility methods to evaluate its integrity if the target structures do not qualify to the design code limits, if this option is allowed by the license.

The energy balance method used here to calculate the whip impact load is straightforward, but it results in very large loads. In fact, what we have done is to say that part of the kinetic energy of the whipping pipe is used to rotate the pipe around the hinge, and the rest of the kinetic energy is converted into the force deflecting the target elastically ($E = {}^1\!/_2\ k\,x^2$ and $F = k\,x$). This approach is quick, but it ignores the energy of deformation of the whipping pipe at the point of impact and the plastic energy absorption of the target. A more realistic but more time-consuming solution is obtained using explicit time-history FEA.

Case Study 5: Temporary Load on Grating

A heavy duty welded steel grating was designed for a live load of 200 psf (1.4 psi). Maintenance requested Engineering to assess whether the grating can sustain a temporary 2000-lb load distributed over a circle 20 in in diameter (Figure 5.20). The live load will therefore be 2000 lb/ 314.2 in^2 = 6.4 > 1.4 psi design. The load will be removed before the end of the outage, before return to power operation.

Before we start solving this problem, let us pay attention to the fact that there are many different types of welded steel gratings, depending on the material (carbon steel, stainless steel, aluminum, or fiberglass), shape, size, and spacing of the bearing bars and the cross-rods supported off the bearing bars, and the finish of the surface (serrated, plain, grip, etc.). Then, there is the method by which the bars and the cross-rods are welded or forged together. Then, there is the means by which panels of grating are joined together to form a floor, and how they are field cutouts. Finally, there is the means by which the grating is secured to the building structures, i.e., the boundary conditions.

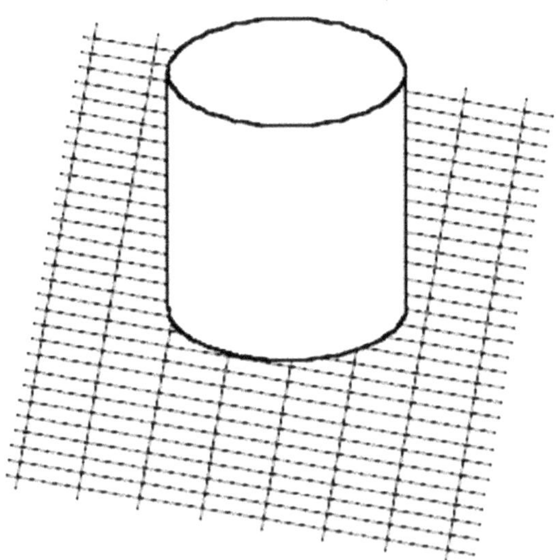

Figure 5.20 Load on grating.

In light of all this, the first thing to do in this case is to retrieve the design drawings and design basis for the grating and perform a filed walkdown to verify (1) that the installed configuration matches the design data, (2) the condition of the grating, for evidence of damage, missing bolts connecting panels, conditions of the attachments to the structure, etc. and (3) the existing dead load on the grating compared to the design load and margins.

Then, once the grating design is retrieved, you will find the type of grating and the applicable catalog, and from the catalog you will obtain: (1) the weight, the safe uniform load (psf) with its maximum deflection, (2) the safe concentrated load (lb/foot of grating width, for a given grating type, with a given width and length) with its deflection, (3) the design-basis maximum stress in the grating elements, which is in the range of 15–20 ksi, depending on the grating material and joints, and (4) the section modulus per foot width.

Given: According to the manufacturer engineering catalog, the grating is ASTM A1011 carbon steel with an allowable stress $F_{allowable} = 18$ ksi, $E = 29{,}000$ ksi, the bars are spaced $a = 1(3/16)$ in apart $= 1.1875$ in, they have a moment of inertia $I_{bar} = 0.178$ in^4 and a section modulus $S_{bar} = 0.1582$ in^3. The span of grating between supports is $L = 3' - 6'' = 42$ in. The normal allowable deflection for grating is $^1/_4$ in for normal traffic. A larger deflection can be permitted safely, but in our case we will limit ourselves to the $^1/_4$ in

The walkdown inspection confirmed the design data and verified that the span is simply supported, with good engagement at the edges, without a risk of slippage of the ends.

The bending moment applied to the grating assembly is

$$M_{bar} = \frac{P(2\,L - D)}{8}\,\frac{a}{D}$$

Note that $D/a = 20$ in$/1.1875$ in $= 16.8$ is the number of bars under the 20-in-diameter load. Substituting $P = 2000$ lb, $L = 42$ in, $D = 20$ in, $a = 1.1875$ in, we obtain $M_b = 950$ in lb applied moment per bar. To limit the bending stress to 18 ksi, the required section modulus of a bar must therefore be larger than

$$S_{bar} = \frac{M_b}{F_{allowable}} = 0.053 \text{ in}^3 > 0.1582 \text{ in}^3 \rightarrow \textit{bar qualifes for bending stress}$$

The deflection of the grating under this 2000-lb load will be

$$\Delta = \frac{P\left[\left(\frac{a}{2}\right)^3 + L^3 - \frac{a^2 L}{2}\right]}{48 \times E \times I_{bar} \times D/a} < 0.25\ \text{in} \rightarrow \textit{bar qualifies for deflection}$$

In summary, the load of 2000 lb distributed over a 20-in-diameter footprint meets the bending stress and deflection criteria with margin. The load will be permitted provided it is placed slowly on the grating without drop. If the load was to be dropped, then we would have to calculate the load on the grating using an energy balance approach. If the weight is P and the drop height H, and if the grating responds elastically with a stiffness k, the drop will deflect the grating an amount Δ_{drop} and the drop impact load P_{drop} will be larger than the static load P, then using the energy balance method, equating the potential energy ($P \times H$) to the grating deflection energy ($^1/_2\ k\,\Delta^2_{drop}$), we obtain

$$\Delta_{drop} = \sqrt{\frac{2\ PH}{k}}$$

$$P_{drop} = k \times \Delta_{drop} = \sqrt{2\ kPH} > P$$

Depending on the shape of the dropped object, we would also have to address the possibility of puncturing the grating.

5.7 SCAFFOLD

First, should we have called this section scaffold or scaffolding?

According to the dictionary, the scaffold is the assembly and the scaffolding are the parts (poles, braces, etc.) that form the scaffold.

What is the purpose of scaffolds in nuclear power plants?

A nuclear power plant is made of large compartments—some are tens of feet tall. To reach equipment at the higher elevations, it is then necessary to build scaffold. They are used to access SSCs for inspection, maintenance, repairs, or modifications.

What do scaffolds in nuclear power plants look like?

Scaffolds in a nuclear power plant are mostly supported scaffolds, as opposed to suspended scaffolds. They are similar to other industrial scaffolds with two important differences: (1) If they are long-term scaffolds, they have to be designed and installed to withstand large dynamic loads, as will be discussed later, and (2) they are in pristine condition, with excellent materials, parts, and workmanship, regularly inspected and maintained.

What are the key elements of a scaffold?

Scaffolds members are illustrated in Figure 5.21 and include the following key members:

- Vertical posts (or poles) are installed plumb and rigidly braced. They are the primary load bearing column members.
- Diagonal cross-bracing is installed along the face and on the sides.
- Runners are horizontal members, placed along the faces, on the inside and outside, at equal heights. The bottom runner is normally as close as possible to the base, at the floor.
- Bearers are horizontal side members, placed on the sides between posts, and they bear on the runners.

What governs the design of a scaffold?

First are the industrial safety requirements: OSHA 29CFR1926 Subpart L "Scaffolds." Valuable information can also be obtained from the UK's National Access and Scaffolding Association. Then, in the case of scaffold installed in nuclear power plants there are material restrictions, and the seismic interaction (SC-II/I) loads and, outdoor, the large wind loads that must be taken into consideration.

There are other unique requirements for scaffolds installed in nuclear power plants:

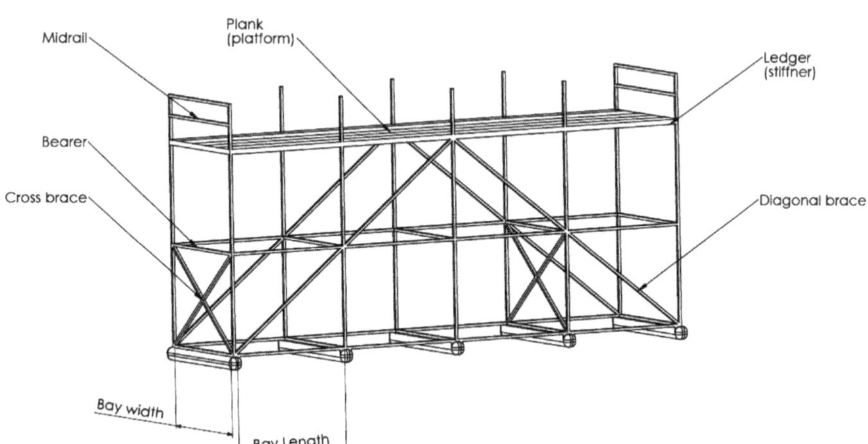

Figure 5.21 General scaffold nomenclature.

- Scaffolds are erected by workers who are experienced and qualified to work in the nuclear power plant environment and in close communication with plant staff.
- Scaffold design and erection must follow ALARA procedures, such as addressed in EPRI "Scaffold Program Optimization and Dose Reduction Guide" 1021102, as well as regulations in 10CFR20, Section 20.1101(b), and NRC Inspection Module 71124.02, "Occupational ALARA Planning and Controls."
- Laser scanning the area to optimize the scaffold design has proved to be a valuable tool for reducing worker exposure and for preventing erection errors and rework.
- If the scaffold will be left in place during a system or plant operation, it may have to be seismically qualified to prevent adverse seismic interactions. Tools must be secured, and nets may have to be used to prevent falling objects.
- It is best to minimize field welding when installing scaffolds, using instead qualified tie wires or clamps.
- Scaffolding must exclude materials not permitted inside containment.

Are scaffolds temporary installations that must be removed once the system is in service?

Much of the scaffold is installed for one outage. But in some cases, the scaffold is used for long term, left in place for multiple outages, to provide access for equipment monitoring and inspection (Figures 5.22, 5.23 and 5.25). For example, permanent scaffold around steam generators or valves saves significant time and dose exposure during in-service inspection and testing. The permanent scaffold also reduces the need for repeated erection and removal, with its risks of mishaps. Permanent scaffold installations would typically be practical in tight locations where the building of permanent platform steel may not be feasible. However, as we will see, if the scaffold has to stay in place for a while, it must be qualified as if it was a welded or bolted structural frame, and this is not always feasible.

More generally, the matter of temporary versus permanent modifications, which includes scaffolding, is addressed in the following documents:

- 10 CFR 50 Appendix B, Criterion V, "Instructions, Procedures and Drawings," which requires procedural controls for activities such as the erection of scaffolding
- Regulatory Guide 1.187, "Guidance for Implementation of 10 CFR 50.59, Changes, Tests, and Experiments"

Figure 5.22 Example of scaffolding in a nuclear power plant.

- NEI 96-07, "Guidelines for 10 CFR 50.59 Evaluations," which addresses temporary changes not related to maintenance

Regarding temporary maintenance modifications, 10 CFR 50.59 evaluations are generally not required provided that (1) the temporary

Figure 5.23 The scaffold must not impair operation of components.

alterations are not in effect longer than 90 days at power and (2) the SSCs are restored to their design condition at the end of the maintenance activity.

For long-term service, can you elaborate on the requirements that must be met?

- The scaffold should be installed to design drawings and erected in accordance with controlling procedures. The drawings and procedures provide restrictions in terms of acceptable bracing, redundancy, and required clearances between scaffold and plant SSCs.
- The scaffolding parts should be connected so that they are prevented from sliding or deforming in a seismic event.
- The scaffold installation should be based on engineering calculations, generated and maintained as part of the plant permanent records. The building steel and concrete to which the scaffold is attached should also be qualified for the additional weight and loads.
- The scaffold design qualification should address normal and accident loads. Accident loads include seismic, pipe break (whip, jet, subcompartment pressurization, ambient temperature), and wind and tornado if outdoor. Pipe whip and tornado loads are more than a scaffold can sustain, so these loads must be prevented. If the scaffold does not qualify under the postulated loads, then it is considered to fail, and in some cases to become a missile. If the interaction effects of the scaffold failure on nearby SSCs are not acceptable, then it cannot be used for long term.
- The scaffold should meet the guidelines provided in OSHA 3150 Information Booklet "Scaffolds Use in the Construction Industry."
- Long-term scaffold should be made of new materials or used materials that are free of defects (nicks, cuts, bends, etc.) and are inspected for good workmanship. Intermixing of component parts from different manufacturers is to be avoided.
- Once erected, the long-term scaffold should be labeled (1) to reflect its long-term permanent status, (2) to require approval prior to dismantling or alteration, and (3) to state safety precautions.
- The long-term scaffold is a design feature of the plant, reflected in drawings, and future changes to existing SSCs should account for the presence of the scaffold framework.
- Scaffold inspections, under the direction of Engineering and plant safety, should be performed prior to the start and at the end of an outage to verify that the scaffold has not been damaged or altered and is fit for use.

- Temporary items mounted on a long-term scaffold, such as tie wire, rubber mats, and duct tape, should be removed from the permanently installed scaffold prior to plant operation.
- The installation should be such that it does not impair the operation of the plant SSCs such as valve stem operation, sprinkler heads effectiveness, crane travel path, access to stairways, operation of electrically energized equipment. In one case, an MSIV valve was found to not fully close due to scaffold decking interfering with the travel of the valve actuator preventing the MSIV from fully closing (IN 07-29). Some plants require a clearance of several inches between scaffold and safety-related components.
- The scaffold should not adversely affect maintenance activities such as impeding access to a manhole. The installation should not result in the damage of sensitive equipment inadvertently.

Can scaffolds survive a seismic event as large as a nuclear power plant's SSE?

Scaffolds can survive earthquakes without becoming a seismic interaction with safety-related SSCs if they are designed, analyzed, qualified, and constructed to seismic criteria.

Scaffolds have been seismically tested on shake tables to determine their failure modes and to develop devices and criteria for their seismic qualification (Figure 5.24). In one particular series of tests conducted in the 1990s, a high-quality scaffold structure from a particular manufacturer, bolted at its base to the shake table, was subjected to a series of triaxial shake table tests. The tests started with a random multifrequency test, followed by resonance search tests, then a series of simulated operating basis earthquakes (approximately 5.5 g peak horizontal spectral acceleration at 5% damping) and SSEs (approximately 7.5 g peak horizontal spectral acceleration at 5% damping), and finally tests at the table limit (approximately 10 g peak horizontal spectral acceleration at 5% damping). The lateral natural frequency of the scaffold was approximately 5 Hz. The damage observed consisted of the loosening of one diagonal brace at the end on the 13th test run and the loosening of a kickboard on the 28th test run.

What was learned from tests regarding seismic design of scaffold?

The shake table tests, as well as structural analyses of scaffolds, have taught us the keys to good seismic design of scaffolds:

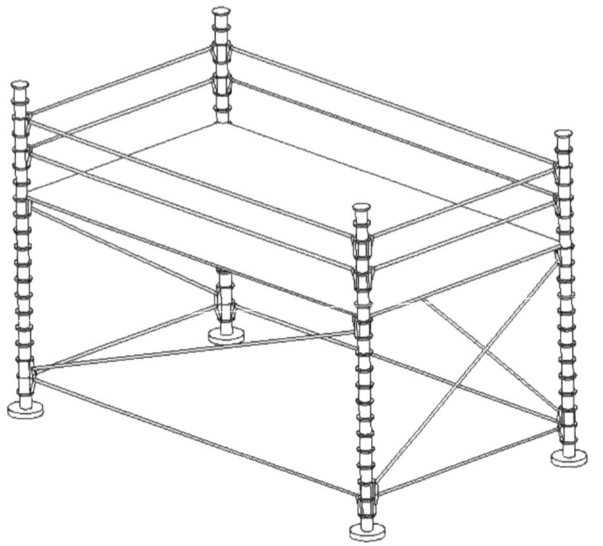

Figure 5.24 General arrangement of a scaffold that was shake table tested.

- The scaffold should be analyzed for the seismic horizontal and lateral loads. The analysis should (1) verify the stability of the scaffold structure, (2) develop the loads at the member joints to select hardware connections or wire ties with sufficient strength, and (3) develop the reaction loads on the scaffold attachment points to the permanent structures.

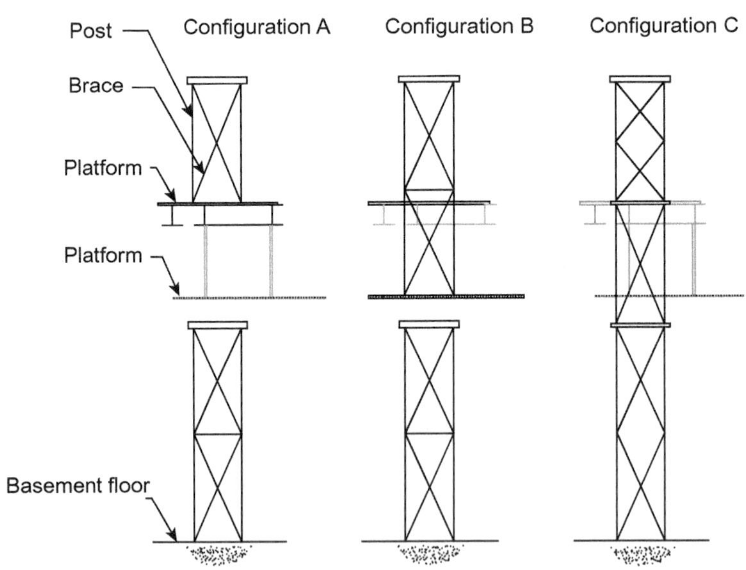

Figure 5.25 Example of a scaffold configuration study.

- Diagonal bracing should be installed along the faces and sides.
- Joints between tubular members should be clamped or tie-wired with connections tested to have sufficient margin against failure or pullout.
- The scaffold structure should be braced against walls or structural beams at a sufficient number of elevations to prevent lateral sway.
- Unless established otherwise by analysis, a minimum clearance should be provided between the scaffold and adjacent safety-related SSCs. Depending on the stiffness of the scaffold and the nearby SSC, this clearance could vary from a few inches to a foot.

How to seismically analyze and qualify a scaffold?

Scaffolds can be analyzed as trusses by hand calculations. They can be modeled and analyzed by finite elements or by using purpose-developed, menu-driven 3D scaffold software, which contains standard elements such as posts, transoms, planks, brackets, etc. The models are then exported to CAD software to develop erection and engineering drawings.

5.8 CASE STUDY

A scaffold in a tight, congested area inside containment has caused several difficulties during its erection. To avoid damage to nearby SSCs, every time the scaffold is erected or dismantled and removed during successive outages, Engineering is asked to evaluate the possibility of leaving the scaffold inside containment as a permanent installation. Following are the steps followed to convert the scaffold into a long-term installation. The steps consist in assembling the design input, defining the loads and load combinations as for any other seismic category I structure, designing the scaffold bracing to sustain these loads without damage, and performing and documenting the analysis and qualification of the braced scaffold.

First, we assemble the design input: a detailed drawing of the scaffold assembly, its support points, and the bill of materials (its member sizes, material, couplings, and catalog data). The scaffold will be supported off four existing structural beams. The loads to be considered are obtained from the FSAR Chapter 3 and are defined as follows: service level $A =$ deadweight of the scaffold and service level $D =$ weight + SSE + pipe break effects. The scaffold will be braced laterally, with the braces bearing against the surrounding walls and clamped to the scaffold members.

Second, we calculate the deadweight load as the total weight and the weight distribution to the four supporting beams. The tube members consist of 1.90″ OD × 0.120″ wall tubing that weighs 2.4 lb/ft. The connector weights are obtained from the manufacturer catalog and weigh between 8 and 10 lb each, depending on the style of connector. The total weight of all the scaffold is 1000 lb, and the maximum weight on the most highly loaded member is 400 lb.

Third, we calculate the seismic load SSE. We select the elevations at which we will place lateral restrains, in the form of compression struts, clamped at one end to the scaffolding and bearing at the other end against the walls, in a pattern that fully restrains each elevation in both lateral directions: east–west (EW) and north–south (NS). We calculate the lateral seismic load by applying the resultant peak spectral acceleration (SRSS of EW and NS) to the tributary scaffold weight at each restrained elevation (500 lb at each of two elevations):

$$F_{lateral} = \sqrt{2\,g^2 + 3\,g^2} \times 500\ \text{lb} \cong 1800\ \text{lb}$$

The maximum vertical spectral acceleration is 0.5 g, so there is no risk of uplift of the scaffold structure. Instead, there is a vertical downward force on the most highly loaded beam equal to gravity 1 g plus the seismic vertical downward acceleration of 0.5 g:

$$F_{down} = (1\ \text{g} + 0.5\ \text{g}) \times 400\ \text{lb} = 600\ \text{lb}$$

We also calculate the seismic load exerted by the scaffolding tubular members on each connecting clamp. The joint load on the highest loaded joint is 90 lb.

Fourth, we calculate the pipe break loads on the scaffold. There are three dynamic loads that have to be considered: pipe whip and jet impingement on the scaffold and subcompartment pressurization. Since the scaffold cannot take a direct pipe whip or jet impingement, it is shielded from these effects by adding barriers as needed. The subcompartment pressurization is obtained from the postaccident mass and energy release environmental analysis of the subcompartment and is provided in the FSAR, Chapter 3.11. In this compartment the maximum pressure is 14 psi for that room. The flow of steam discharging from a break around the 1.90″ OD tubular member is calculated to cause a drag force:

$$F_{drag} = \frac{1}{2} C_D \rho U^2 A$$

Where F_{drag} is the lateral drag force on the tubular member, C_D is the drag coefficient that is small for such a small tube, ρ is the density of the fluid, and A is the cross-section of the tube that intersects the flow. The drag force is a distributed force and is calculated to result in near-zero deflection and stresses of the scaffold members. The walking planks do present a large intersecting area for the drag force, but they will be removed at the end of each outage, so they will not be subjected to the postulated pressurization of the subcompartment.

Fifth, we address the temperature rise in the subcompartment as a result of the pipe break—the thermal growth of the scaffold steel structure. The thermal expansion of the poles is calculated and is verified to be accommodated by the joints.

Sixth, the weight plus SRSS SSE and pipe break loads on the scaffold members and joints are verified to be within the manufacturer limits. The reaction loads on the supporting beam structures are added to the existing loads from the design analysis and verified to be within the design code allowable.

Finally, we document the design, the analysis, and the qualification following the same procedures as other permanent structures.

5.9 COMPONENT CLEARANCES

Why are there clearance requirements in nuclear power plants?

Some areas in a nuclear power plant are, by necessity, congested, which leads in some cases to little space between SSCs, as shown in Figure 5.26.

Figure 5.26 Example of tight clearance near a valve motor operator.

The purpose of having clearance requirements between commodities, including temporary installations, is to prevent adverse interaction from occurring between these commodities. These could be accidental interactions like bumps during maintenance or interactions during normal operation from heat-up expansion or interactions during postulated accidents such as seismically induced swaying.

Are these spatial interferences due to normal operation and seismic interactions are the only concern?

There are other concerns with SSCs being too close to each other, for example, accessibility for inspection, operation, and maintenance. Another consideration is that hot or warm equipment must be kept clear to provide for sufficient ventilation cooling. For example, clearances are evaluated by electrical engineering and fire protection engineering to verify that safety-related cables are shielded from the radiant heat of nearby hot piping.

In a few cases, in what seems like a tight arrangement, such as two pumps sharing a common skid, the proximity may be acceptable and is accounted for in the design process.

How are the commodities clearance requirements implemented in practice?

Design loads that can cause large movements are analyzed carefully as part of the original plant design, and the movements are compared with the available clearances, with margin added to the space between commodities.

During preoperational testing, the heat-up expansion of systems is monitored to verify that there are no interferences. Then, once the plant is in operation, operator rounds check for abnormal condition, including interferences. Plant procedures require several design and inspection precautions, which include the following:

- Sufficient distance between SSCs during erection, with consideration of expansion and swing
- Sufficient distance between Class 1E electrical equipment and heat sources
- Sufficient distance to not impair the function and spray zone of sprinklers

These precautions are achieved by formal reviews of conceptual designs for clearance requirements. Keep in mind that the clearance between two components has to take into account insulation, protective jackets, and shielding.

Are there any special considerations for hung systems, like light fixtures or flexibly suspended pipes?

SSCs that are by design flexible, such as hung light fixtures or pipe supported by rod hangers, must be evaluated for the zone of influence of their swinging movements. If the interaction is not acceptable, the potential for swinging will be removed, most often by bracing the component. For example, in the control room, things like suspended ceilings and light fixtures are restrained from swinging or falling, sometimes using aircraft cable. These aircraft cables are also used to secure pipes to walls or structures and prevent them from swinging or falling, as shown in Figure 5.27.

How are clearances verified in the field?

Verifying commodity clearances on an ongoing basis is primarily the responsibility of an assigned department such as Quality Control, Engineering, or Maintenance, but it is also everyone's responsibility to flag potential clearance issues. In each case the condition is evaluated. If the proximity is unacceptable, or in case of doubt, corrective measures are taken.

Which SSCs are affected by insufficient clearance?

There are some commodities that are more often flagged for clearance concerns such as

- distribution systems (piping and tubing and their supports; conduit, cable trays, HVAC duct, and their supports)

Figure 5.27 Fire protection pipe secured by aircraft cable.

- platforms, ladders, and handrails
- electrical components because of the need for adequate ventilation and because most electrical components cannot withstand rattle impact from nearby commodities
- scaffold and shielding

CHAPTER 6

Materials and Aging Mechanisms

Chapter 6 covers, in its first part, the materials commonly used in nuclear power plants and the exacting material quality controls that apply. The second part addresses the key corrosion issues that have surfaced and have been tackled in our industry in the recent years.

6.1 COMMON MATERIALS

What materials are most commonly used in a nuclear power plant?

The materials used in a mechanical equipment and systems in a nuclear power plant are primarily metallic materials that comply to a material specification, with sometimes supplementary requirements. The nuclear industry uses, for the most part, carbon and low-alloy steels, stainless steels, and Ni alloys.

What are examples of differences between nuclear versus nonnuclear materials?

Some key differences between nuclear power plant materials and those in other industries are:

- Materials for safety-related applications must be procured from qualified suppliers who must implement a rigorous verification and quality control process for their materials, with serious penalties if these controls are not followed or are tampered with. These quality assurance requirements are captured in 10CFR50 Appendix B and in ASME NQA-1.
- Materials used must be traceable from the point of origin to the point of installation. As an example, ASME III NB-4000 (Class 1 materials) requires that each container of weld wire shall be traceable to the lot, production period, and even the shift used to make the wire. This is true for the newly constructed plant and remains true throughout the operating life of the plant, for 60 or more years.
- Materials for pressure-retaining components and weld must comply with ASME II specifications. In nonnuclear plants this is true for boilers and pressure vessels, but in a nuclear plant this remains true for piping, tubing, pumps, and valves.

Nuclear Power Plant Safety and Mechanical Integrity
ISBN 978-0-12-417248-7

- ASME Safety Class 1 materials (the reactor coolant system) have exacting supplementary material requirements. For example, with few exceptions, plates used to make Class 1 vessels must be 100% volumetrically examined by straight beam and angle beam UT.

The quality, pedigree of supplied materials is closely monitored?

Yes, very closely, and this is to make sure that what comes on site is what is called out in the design, that there are no errors of inattention or counterfeits. There have been instances of noncompliant materials supplied to plants, which were detected and rejected thanks to this close quality control process. The NRC has issued Generic Letter (GL) 89-02, "Actions to Improve the Detection of Counterfeit and Fraudulently Marketed Products" to emphasize the characteristics of an effective procurement process that would catch flawed or counterfeited materials:

- The involvement of technical engineering staff in preparing the procurement specification and then in verifying the quality of deliverables
- The necessity of technically robust inspection of suppliers, receipt inspections, and testing programs
- Thorough engineering-based programs for review, testing, and dedication of items that were manufactured by noncertified suppliers, what is referred to as commercial-grade dedication

Other references of interest to this question of quality control of procured items include:

- IN 89-70 "Possible Indications of Misrepresented Vendor Products"
- IN 2008-04, "Counterfeit Parts Supplied to Nuclear Power Plants"
- "Safety Culture Policy Statement," June 14, 2011, NUREG/BR-0500, "Safety Culture Policy Statement"
- Regulatory Issue Summary 2012-01, "Availability of Safety Culture Policy Statement"

What are ASME II specifications, and how different are they from the usual ASTM specifications?

ASME material specifications are material specifications listed in ASME II. ASME II materials are denoted with the letter S in front of the specification number. For example, an ASTM A-106 pipe that meets ASME II will be labeled and marked SA-106. They are similar to ASTM material specifications, and many are identical. But in some cases there are differences between ASTM and ASME II material specifications, which are stated in the front of each ASME II specification, in a statement such as "identical to ASTM, except for ...."

What information other than material specifications is provided in ASME II?

The ASME Boiler and Pressure Vessel Code Section II is entitled Materials, and contains four parts, A through D:

- Part A contains the material specifications for ferrous materials.
- Part B contains the material specifications for nonferrous materials.
- Part C contains the material specifications for welding filler metals.
- Part D contains the properties of the materials. Within Part D, there are three subparts: (1) Subpart 1 includes the allowable stresses for all materials as a function of temperature, which is an essential input to design. It also contains the maximum temperature limit for each material. (2) Subpart 2 provides the modulus of elasticity and the coefficient of thermal expansion of ASME materials. (3) Subpart 3 provides the charts needed for external pressure design.

So, for safety class, the pressure boundary components and their support must have an ASME II pedigree?

Yes, but it depends on the plant vintage and license. Some of the older plants may have been constructed to ASME VIII or ASME B31.1 with ASTM materials. In all cases, the requirements for materials must conform to the plant's final safety analysis report. We must not forget that there are material requirements in ASME III itself (such as in ASME III Appendix I), and, for nonsafety systems, there are material requirements and limitations in the corresponding code and standard, for example, ASME B31.1 for piping and ASME B16.34 for valves.

What materials are most commonly used?

The hierarchy of materials used starts with carbon steel and low-alloy steels (Cr-Mo steels). Most plant systems (piping, vessels, pumps, valves, cable trays, ducts, etc.) as well as structural members are made of carbon steel. Then, stainless steel is used in the more corrosive services such as the primary system of pressurized water reactors (PWRs), which operates with boric acid at about 650 °F. The preference is for the L grades (304L or 316L) with the lower carbon content, which reduces the risk of chromium carbide precipitation (sensitization) of the stainless steel during welding. One level above stainless steel are nickel alloys, which are often used as buttering for welds between carbon and stainless steel. One of the advantages of Ni alloys is that their coefficient of thermal expansion is similar to that of carbon steel. However, nickel alloys are not a panacea; for example, alloy 600 (and its weld metals alloys 82 and 182) has proven to be susceptible to primary water stress corrosion cracking (PWSCC) in long-term service in a PWR primary water environment and has been replaced by alloy 690 (and its weld metal alloys 52 and 152). On the secondary side such as in the condenser, where steam is condensed back into water, the heat exchanger and condenser tubes can be made of copper alloys or titanium alloys. Brittle materials such as cast iron are avoided in safety-related systems.

What are the common materials for supports?

Many supports are fabricated from carbon steel, which (for safety-related systems and equipment supports) have to be procured to an ASME II specification (safety class) or ASTM (nonsafety class). The safety class supports are designed to ASME III NF in the newer plants. Common materials for support parts include SA-36 (structural carbon steel), SA-240 (stainless-steel sheet and strips), SA-500 (structural tubing), SA-570 (sheet and strips) now SA-1011, SA-572 (high-strength low alloy), SA-653 (galvanized sheet), and SA-992 (structural steel shapes). If a support component has to be welded to the pipe, to make an anchor or to be used as a rectangular lug or round trunnion, then the material has to be compatible with the pipe, for example, SA-240 strips or sheets on stainless-steel pipe or SB-574 on Ni alloys. SA-325 is commonly used as carbon steel bolt material for supports (high-strength structural bolts). Welds in carbon steel supports are typically made of E70XX filler metal. A good reference list of materials for supports can be found in Standard Practice MSS-SP-58.

In addition to corrosion resistance and radiation resistance, what other material considerations apply for nuclear plant materials?

From a design point of view, the material must have sufficient strength (yield strength, ultimate strength, and elongation at rupture) at the operating temperature, as well as good fatigue resistance, and good toughness (Charpy notch toughness or fracture toughness).

From a fabrication point of view, there is of course the need for good weldability.

What are unique requirements for nonmetallic materials?

For resistance to fire and smoke and for low toxicity, there is a need for halogen-free electrical cables and even duct tape. There is, for example, a nuclear-grade duct tape, which has low halogens and which leaves little or no residue. These types of limitations on nonmetallic materials are managed through material exclusion areas in nuclear power plants, where any material that enters is preverified, screened, and controlled.

Insulation has to be resistant to impingement by high-pressure water or steam jets from a postulated pipe break. The concern is that the jet would tear the insulation, forming debris (particulates, fibers, reflective metal insulation foils) that can accumulate and clog recirculation sumps (addressed in NEI 04-07, NRC Generic Letter 2004-02, NUREG/CR-6808). This leads to the use of reflective metal insulation, which has been tested for its resistance to jet impingement.

6.2 AGING MECHANISMS

What damage mechanisms are unique to nuclear power plants?

A nuclear power plant circulates water and steam in metallic pipes and components. The fluid systems consist of (1) the primary water (the water that circulates inside the reactor) and (2) the water and steam in other systems, such as the chemical and volume control system, the safety injection system, the residual heat removal system, the spent fuel cooling system, contain boric acid.

The plants are subjected to the classic corrosion mechanisms for carbon steel and stainless steel in a water or steam service at temperatures ranging from ambient to 650 °F. The material damage mechanisms in a nuclear power plant can best be understood by the plant engineer if we group them into three categories:

- Wall thinning, such as galvanic effects, microbiologically induced corrosion, flow-accelerated corrosion (FAC), soil-side corrosion in buried pipes, and erosion.
- Cracking, such as stress corrosion cracking (SCC), in particular PWSCC, and fatigue, either low cycle or high cycle (such as vibration induced).
- Embrittlement, in the form of loss of physical or mechanical properties and, in the case of neutron embrittlement, swelling.

The close chemistry controls and the strict in-service inspection programs implemented in nuclear power plants significantly reduce these detrimental effects.

Because of their lower operating temperatures, nuclear power plants are spared the creep mechanisms of fossil power plants.

Where to start on a topic as vast and complex as damage mechanisms in a nuclear power plant?

It is a difficult subject, but, because this is the nuclear power industry, these issues are well documented, published, and well structured, thanks to the efforts of industry engineers, the NRC, EPRI, INPO, NEI, IEAE, and others. The question of damage mechanisms came to the forefront when nuclear plants applied for life extension to extend their operation from their original 40 years to an extended life of 60 years or more. The key questions were what would be the damage mechanisms in the aging plants, what have we learned in 40 years, and how can this knowledge be used to predict life to 60 years and beyond. To tackle this question, and a good starting point to address damage mechanisms, the NRC developed and published NUREG-1801 "Generic Aging Lessons Learned (GALL) Report." This report was developed as a technical reference to NUREG-1800, "Standard Review Plan for Review of License Renewal Applications for Nuclear Power Plants" (SRP-LR) for license renewals under 10 CFR Part 54, "Requirements for Renewal of Operating Licenses for Nuclear Power Plants."

- NUREG-1801 Volume 1 (approximately 100 pages) is a tabular list of aging mechanisms for (1) the reactor coolant system, (2) the auxiliary systems, and (3) the steam and power conversion systems. The columns are (1) the type of reactor (PWR or BWR), (2) the type of material and component affected, (3) the aging mechanism, (4) the aging management program (AMP) under which it is addressed, (5) whether further evaluation is recommended, and (6) the cross-reference to the applicable section in Volume 2.
- NUREG-1801 Volume 2 (approximately 900 pages) systematically reviews every system in a PWR and BWR, and identifies the corresponding aging mechanism.

What role has life extension played in improving our understanding of damage mechanisms?

The licensing of nuclear power plants to extend the original 40-year design life by an additional 20 years has been an incentive for a systematic and in-depth assessment of aging mechanisms. This was a formal requirement under the license renewal regulation, under 10 CFR 54.21, which states: "Each application must … identify and list those structures and components subject to an aging management review." 10 CFR 54.21 goes on to require that the license extension must include analyses of time-limited aging mechanisms, and that the applicant must demonstrate that "(1) the analyses [of time-limited aging mechanisms must] remain valid for the period of extended operation; (2) the analyses have been projected to the end of the period of extended operation; or (3) The effects of aging on the intended function(s) will be adequately managed for the period of extended operation."

What are the systems covered in NUREG-1801?

The list of systems covered in NUREG-1801 is a complete compilation of the systems in a PWR and BWR; in the practice of plant operations, each one of these systems is assigned a system engineer:

- Containments: PWR containments: concrete containments (reinforced and prestressed), steel containments; BWR containments: Mark I, II and II containments
- Structures: BWR reactor building, PWR shield building, control room/building; BWR reactor building with steel superstructure; auxiliary building, diesel generator building, radwaste building, turbine building, switchgear room, AFW; pumphouse, utility/piping tunnels yard structures such as AFW pumphouse, utility/piping tunnels, security/lighting; poles, manholes, duct banks; SBO structures such as transmission towers, start-up towers circuit breaker foundation, electrical enclosure; containment internal structures, excluding refueling canal; fuel storage facility, refueling canal; water-control structures; concrete tanks and missile barriers; steel tanks and missile barriers; BWR unit vent stack
- Component supports: supports for ASME piping and components; supports for cable trays, conduit, HVAC ducts, instrument tubing, non-ASME piping and components; anchorage of racks, panels, cabinets, and enclosures for electrical equipment and instrumentation; supports for emergency diesel generator (EDG), HVAC system components, and other mechanical equipment; supports for platforms, pipe whip restraints, jet impingement shields, masonry walls, and other miscellaneous structures
- Reactor vessel, internals, and reactor coolant system: BWR and PWR reactor vessels, reactor internals by manufacturer, reactor coolant pressure boundary (BWR); reactor coolant system and connected lines (PWR); steam generator (recirculating); steam generator (once-through)

- Engineered safety features: containment spray system (PWR); standby gas treatment system (BWR); containment isolation components; emergency core cooling system (PWR and BWR)
- Electrical components: equipment, electrical cables and connections not qualified to 10 CFR 50.49 environmental qualification (EQ), equipment qualified to 10 CFR 50.49 EQ
- Auxiliary systems: new fuel storage, spent fuel storage, spent fuel pool cooling and cleanup (PWR and BWR), suppression pool cleanup system (BWR), overhead heavy load and light load (related to refueling) handling systems, open-cycle cooling water system (service water system), closed-cycle cooling water system, ultimate heat sink, compressed air system, chemical and volume control system (PWR), standby liquid control system (BWR), reactor water cleanup system (BWR), shutdown cooling system (older BWR)
- Ventilation systems: control room area ventilation system, auxiliary and radwaste area ventilation system, primary containment heating and ventilation system, diesel generator building ventilation system
- Fire protection systems: diesel fuel systems, diesel fuel oil system, EDG system
- Steam and power conversion system: steam turbine system, main steam system (PWR and BWR), extraction steam system, feedwater system (PWR and BWR), condensate system, steam generator blowdown system (PWR), auxiliary feedwater system (PWR)

What are the aging analyses covered under 10 CFR 54.21(c)(1)(iii)?

As mentioned earlier, license renewal applications had to include the study and actions to prevent and mitigate aging mechanisms. The three programs listed in NUREG-1800 as being directly covered under 10 CFR 54.21 are as follows:

- Metal Fatigue of Reactor Coolant Pressure Boundary
- Concrete Containment Tendon Prestress
- EQ of Electrical Components

How is the first issue, metal fatigue of the reactor coolant pressure boundary, addressed in design?

A key aspect of safe operation is to prevent fatigue cracking of the reactor coolant system base metal and welds. These are subjected to mechanical and thermal load cycles while in service, and they have to be analyzed and qualified for a fatigue life that will be longer than the design life of the system. This fatigue analysis was done in the past following the fatigue design rules of ASME III Division 1, Class 1 (subsection NB), which consists of a few key steps, described in a simplified form as follows:

- At the design stage, the system transients are tabulated in the form of pressures (P), flow rates (Q), and temperatures (T) as a function of time $P_i(t)$, $Q_i(t)$, and $T_i(t)$ for each operating mode or transient i.
- These $P_i(t)$, $Q_i(t)$, and $T_i(t)$ transients are applied to a heat transfer and finite element model of each reactor coolant system component.
- The cyclic stresses, σ_i, in each component of the reactor coolant system are then calculated for each transient $P_i(t)$, $Q_i(t)$, and $T_i(t)$.
- The cyclic stresses, σ_i, for each transient $P_i(t)$, $Q_i(t)$, and $T_i(t)$ are then converted into principal stresses and then into the maximum shear stress, called the stress intensity range, S_i.
- For each stress intensity range, S_i, the fatigue curves contained in ASME III Appendix I will give us a number of allowable cycles, N_i.
- Then, we calculate the ratio $u_i = n_i/N_i$ of the predicted number of cycles n_i for 40 years (or 60 years) of operation to the allowed number of cycles N_i from the ASME III Appendix I fatigue curve.
- We add this ratio u_i for all the cycles $P_i(T)$, $Q_i(t)$, and $T_i(t)$ over the 40 year design life as the total usage factor $U = \Sigma u_i = \Sigma n_i/N_i$ and verify that the sum U is below 1.0.

How has this changed over the years?

A concern has surfaced in the last 10 years or so that has been labeled environmental fatigue. In simple terms, the concern is that the design fatigue curves in ASME III Appendix I, which we used to obtain the allowable number of cycles N_i, were based on fatigue tests of smooth bar specimen in a laboratory, in air, not in the primary water environment at operating temperature of 550–650 °F, with oxygenated water, boric acid, and at variable strain rates. But the ASME III Appendix I fatigue curves include a safety factor of 20 on cycles and 2 on stress, so another way to phrase the concern is to ask whether these safety factors are sufficient to compensate for the corrosive environment of the primary system water, with still an acceptably large safety factor.

Formulas for calculating the environmental fatigue penalty factors were provided in NUREG/CR-6583 "Effects of LWR Coolant Environments on Fatigue Design Curves of Carbon and Low-Alloy Steels" March 1998, for carbon and low-alloy steels and in NUREG/CR-5704 "Effects of LWR Coolant Environments on Fatigue Design Curves of

Austenitic Stainless Steels," April 1999, for austenitic stainless steels. These environmental fatigue penalty factors were then applied to quantify the new, reduced fatigue life for critical components; this was done in NUREG/CR-6260 "Application of NUREG/CR-5999 Interim Fatigue Curves to Selected Nuclear Power Plant Components," March 1995.

An important aspect of correctly predicting fatigue life is to get a good handle on the transients $P_i(T)$, $Q_i(t)$, and $T_i(t)$ themselves. While this was difficult to do 40 years ago when the rules were being applied to design many of the operating plants in the US, this is being done today thanks to cycle monitoring, which permits the collection of actual plant heat-up and cooldown records.

What is the concern with the second issue, concrete containment tendon prestress?

The level of prestress in concrete tendons for containments is inspected in accordance with ASME XI Subsection IWL and 10 CFR 50.55a(b)(2)(ix) or (viii) "*Examination of metal containments and the liners of concrete containments.*" The results are trended following the guidance of RG 1.35.1 "Determining Prestressing Forces for Inspection of Prestress Concrete Containments," and Information Notice IN 99-10 "Degradation of Prestressing Tendon Systems in Prestressed Concrete Containments." The objective is to keep the tendon prestress trend line above a predicted lower limit and a minimum required prestressing force value (MRV).

What is the concern with the third issue, EQ of electrical components?

10 CFR Part 50, Appendix A, Criterion 4, and 10 CFR 50.49 require that essential electrical components that must operate after a postulated break be qualified for the harsh environment caused by the break, for example, the humidity, the high temperature, the potential spray and flooding, and the boric acid environment that would result from a guillotine break of the reactor coolant loop. Recall that these environmental effects are not excluded even if it is established that the RCL is qualified for leak before break. These qualifications are accomplished through the EQ program. 10 CFR 50.49 requires that the effects of significant aging mechanisms be addressed as part of EQ. Simply put, will the electrical equipment operate in the harsh environment of a break in a flashing water or steam line at year 39, and, after life extension, at year 59?

The question of EQ dates back to the 1970s, culminating in 1979 with the issue of Bulletin 79-01 "Environmental Qualification of Class IE Equipment," which requested plants to review the adequacy of their EQ records for Class 1E electrical equipment, to verify that they would function in the harsh environments that would result from a postulated pipe break.

Since then, guidance for EQ has been established in several documents:

- The November 1979 NRC memorandum "Guidelines for Evaluating Environmental Qualification of Class 1E Electrical Equipment in Operating Reactors," which provides good technical guidance related to EQ, still valid 35 years later
- Standard Review Plan Section 3.11 "Environmental Qualification of Mechanical and Electrical Equipment"
- Regulatory Guide 1.89 "Environmental Qualification of Certain Electric Equipment Important to Safety for Nuclear Power Plants"
- 10 CFR 54.21(c)(1)(iii) contents and application for life extension
- IEEE standards 317 (penetrations), 323 (Class 1E), 334 (motors), 382 (electrical valve operators), 383 (cables and field splices), 535 (lead storage batteries), 572 (Class 1E connections)
- NUREG-0588 (EQ), 0737 (TMI action plan)
- RG 1.63 (penetrations), 1.73 (electrical valve operators), 1.89 (electrical equipment), 1.97 (environment instrumentation), 1.131 (cables and field splices), 1.156 (connection assemblies), 1.158 (lead storage batteries), 1.160 (maintenance), 1.209 (computer-based instrumentation)

What are the damage mechanisms covered under the AMPs?

The damage mechanisms covered under the AMPs are a good snapshot of the material issues faced by nuclear power plants. These issues evolve as old ones are resolved, new ones appear, or priorities change; based on NUREG-1801, the AMPs include the following:

- Water chemistry
- Reactor head closure studs
- BWR vessel ID attachment welds
- BWR feedwater nozzle
- BWR control rod drive return line nozzle
- BWR SSC
- BWR penetrations
- BWR vessel internals
- Boric acid corrosion

- Nickel alloy nozzles and penetrations
- Nickel alloy penetration nozzles welded to the upper reactor vessel closure heads of pressurized water reactors
- Thermal aging embrittlement of cast austenitic stainless steel (CASS)
- Thermal aging and neutron irradiation embrittlement of CASS
- Loose part monitoring
- Neutron noise monitoring
- PWR vessel internals
- FAC
- Bolting integrity
- Steam generator tube integrity
- Open-cycle cooling water system
- Closed-cycle cooling water system
- Boraflex monitoring
- Inspection of overhead heavy load and light load
- (Related to refueling) handling systems
- Compressed air monitoring
- BWR reactor water cleanup system
- Fire protection
- Fire water system
- Buried piping and tanks surveillance
- Aboveground steel tanks
- Fuel oil chemistry
- Reactor vessel surveillance
- One-time inspection
- Selective leaching of materials
- Buried piping and tanks inspection
- One-time inspection of ASME Code Class 1 small bore piping
- External surfaces monitoring
- Flux thimble tube inspection
- Inspection of internal surfaces in miscellaneous piping and ducting components
- Lubricating oil analysis
- 10 CFR 50, Appendix J
- Masonry wall program
- Structures monitoring program
- RG 1.127, inspection of water-control structures associated with nuclear power plants
- Protective coating monitoring and maintenance program

- Electrical cables and connections not subject to 10 CFR 50.49 EQ
- Electrical cables and connections not subject to 10 CFR 50.49 EQ requirements used in instrumentation circuits
- Inaccessible medium-voltage cables not subject to 10 CFR 50.49 EQ requirements
- Metal-enclosed buses
- Fuse holders
- Electrical cable connections not subject to 10 CFR 50.49 EQ requirements

Prevention of failure due to age-related damage implies periodic inspections, how are they implemented in nuclear power plants?

The plant programs by which the effects of aging are monitored and trended follow ASMEXI, which defines the periodic inspection program for ASME III Class 1 (IWB), Class 2 (IWC), and Class 3 (IWD). ASME XI applies to pressure-boundary components and their welded attachments (we will refer to IWB or IWC or IWD as IWX). The inspection requirements are specified in Tables IWX-2500-1. The inspections are repeated periodically, in the same sequence as established during the first inspection interval, as much as possible so as to trend the inspection readings. The Section XI ISI programs focus on two general areas:

- The damage mechanisms themselves: wall thinning by corrosion or erosion, cracking by corrosion or fatigue, embrittlement, mechanical damage
- The telltales of the damage such as leaks, settings, clearances, preload, etc.

The ISI are based on three inspection techniques:

- Visual may be VT-1 for imperfections, VT-2 for leakage, and VT-3 for any condition that affects adequacy or operability.
- Surface examinations (excluding VT) are typically based on PT (liquid pentrant) and MT (magnetic particles) and are meant to detect crack-like surface flaws.
- Volumetric examination are typically RT (radiographic), UT (ultrasonic), or ET (eddy current) and are meant to detect flaws within the volume or the inaccessible inner or outer surface of the material.

The inspection of Class 1 components is specified in Table IWB-2500-1 and is subdivided into examination categories:

- Examination category B-B for volumetric examination of circumferential and longitudinal vessel welds

- Examination category B-D for full-penetration welds of nozzles
- Examination category B-E for pressure-retaining partial penetration welds in vessels
- Examination category B-F for pressure-retaining dissimilar metal welds (DMW)
- Examination category B-G for pressure-retaining bolting
- Examination category B-K for integral attachments for vessels
- Examination category B-J for pressure-retaining welds in piping
- Examination category B-L for pressure-retaining welds and base metal in pump casing
- Examination category B-M for pressure-retaining welds and base metal in valve bodies
- Examination category B-N for the interior of reactor vessels and their internals
- Examination category B-O for pressure-retaining welds in control rod housing

The examination of Class 2 components follows Table IWC-2500-1:

- Examination category C-A for pressure-retaining welds in pressure vessels
- Examination category C-F-1 for pressure-retaining welds in austenitic stainless steel or high-alloy piping
- Examination category C-G for all pressure-retaining welds in pumps and valves
- Examination category C-H for all pressure-retaining components

The examination of Class 3 components follows Table IWD-2500-1:

- Examination category D-A for systems in support of reactor shutdown function
- Examination category D-B for systems in support of emergency core cooling, containment heat removal, atmosphere cleanup, and reactor residual heat removal

What is primary water stress corrosion cracking (PWSCC)?

In a PWR, the pressure vessel is made of thick sections of ferritic steel. To protect the ferritic steel from the corrosive effect of the primary water (boric acid at high operating temperature, above 600 °F), the vessel is lined with austenitic stainless steel. The welds between the ferritic steel vessel and the austenitic stainless steel reactor coolant pipe and vessel penetration nozzles, dissimilar metal joints (DMW), are made with Inconel. The Inconel most commonly used in nuclear power plants is Inconel 600 with its weld metal

82 (gas metal arc welding and gas tungsten arc welding) and 182 (shielded metal arc coated electrode). Inconel 600 is a nickel alloy with 72% minimum nickel, 14–17% chromium, and 6–10% iron.

In broad terms, PWSCC is the cracking of Inconel 600 and its weld material in a PWR reactor water environment. This cracking leads to hot (over 600 °F) pressurized (over 2000 psi) primary water leaking through the crack and contacting the carbon steel. The leaking boric acid then causes corrosion of the ferritic steel, in the form of wall loss, material wastage.

So, in summary, we have a PWSCC cracking mechanism of the Inconel, followed by a boric acid corrosion-induced wall loss mechanism of the reactor vessel's ferritic steel. NRC Bulletin 2001-01: Circumferential Cracking of Reactor Pressure Vessel Head Penetration Nozzles provides further details.

What components are susceptible to PWSCC?

All locations containing Inconel 600 in contact with primary water are susceptible to PWR PWSCC. There are several such locations as follows (NRC Regulatory Issue Summary 2003-13):

- The penetrations to the reactor pressure vessel from the top control rod drive mechanisms (CRDM) and from the bottom (bottom-mounted instrument tubes)
- The pressurizer penetrations (heaters, safety valve nozzles, etc.)
- The steam generator (SG) tubes and tubesheets
- The reactor vessel nozzles to the reactor coolant loop piping
- The SG nozzle welds

What causes PWSCC?

The usual suspects in SSC are at play here; they are the overlapping effects of (1) the environment, (2) the material, and (3) the stress. Each of these has been studied in detail by the industry through extensive programs such as EPRI's material reliability programs MRP, which for PWSCC go back over 20 years. Various potential causes for PWSCC have been identified:

- The environment: Regarding the primary water environment, a key contributor is temperature. For example, pressurizer nozzle repairs have failed in approximately 1 year at 650 °F, while a cold-leg pipe instrument nozzle failed in 10 years at 550 °F. It seems that the water chemistry (boron, lithium, and pH) has relatively little effect. However, adjusting hydrogen and oxygen content of the pressurized water has been reported to have some effect on PWSCC crack growth rate. The presence of zinc seemed to help, and some plants have used Zn injections to mitigate PWSCC.

- The material: Regarding the metal itself, the presence of intragranular carbides (carbides within the grain) increases the susceptibility to PWSCC, and therefore a low carbon content is desired. On the other hand, intergranular carbides (carbides at the grain boundary) tend to prevent the occurrence of PWSCC. Favorable conditions to prevent PWSCC are therefore achieved with a large grain size, and slow cooling to maximize carbide precipitation at the grain boundaries. This explains why, for the same material specification, PWSCC has proven to vary from heat to heat. The complex dendritic grain structure of multiple-pass 82/182 welds also has a detrimental effect on crack propagation. Other important parameters are surface finish, roughness of the weld, and grinding (cold working) of the surface. Finally, the chromium content of the nickel alloy is important; a reduction in crack growth rate by a factor of nearly 100 was achieved by going from 14–17% Cr (82/182 welds) to 27–31% Cr (52/152 welds).
- The stress field: Regarding stresses, it appears that PWSCC requires a stress larger than the usual 30 ksi yield strength at temperature. If the components have been designed correctly to meet ASME III stress limits, this level of stress should not occur from applied membrane and bending loads. In this case, the key contributing factor would be the peak stresses and surface residual stresses generated by surface finishing operations that may have been performed after stress-relieving heat treatment, or by as-welded conditions that did not receive a postweld stress-relieving heat treatment.

Would such cracks not be hard to detect under the insulation?

In fact, what was first discovered was boric acid attack of ferritic steel not leaking from PWSC cracks, but leaking from gaskets and seals, back in May 1980 at a US PWR. During a routine pressure test of the reactor coolant system at 180 psi, the insulation started to drip around the reactor coolant pump flange. A number of pump flange studs (ASTM A183 Grade 87, 3.5-inch diameter) were found to be severely corroded. The corrosion was eventually attributed to attack by boric acid from the primary water that had leaked passed the gasket and wetted the bolts (NRC Information Notice 80-27). Following this first event, similar wall thinning from boric acid was identified in bolts and studs at other plants.

In 1987, another PWR detected boron crystals on the reactor vessel head, a sign of leakage of hot primary water. The primary water, which contains boric acid, flashes to steam as it leaks, leaving behind deposits of boric acid crystals. The concentration of boric acid by vaporization of the leaking water accelerates the corrosion of steel. This leak had corroded vessel head bolts and CRDM shroud supports. It was again a case of wall thinning (wastage) of the carbon steel by corrosive attack.

Several cases followed and are documented in NRC Information Notices 80-27, 82-06, 86-108, and 86-108; Supplements 1 and 2; Bulletin 82-02; and Generic Letter 88-05. In 1988, PWSCC of Inconel 600 was identified as the culprit for cracks in pressurizer sleeves (Information Notice 90-10).

Were the late 1980s the first time that cracking of Inconel 600 was an issue?

No, we are sometimes reminded of the work in the late 1950s by H. Coriou and his colleagues in France on SSC of Inconel in high-temperature service. In the 1970s, cracking of Inconel 600 had occurred in SG tubes, with leakage from the primary side (reactor coolant) to the secondary side (feedwater and steam). The first SG replacements were made in the US in the early 1980s, and continued through the 1980s and 1990s. It is through the SG effort that the industry studied replacement materials such as Inconel 690 and 52/152 weldments.

How is the issue of PWSCC solved across the industry?

The NRC and the industry have implemented priority actions to first mitigate, then prevent the PWSCC problem. PWSCC is being solved by a combination of replacement, repairs, and augmented inspections. Because the locations of PWSCC are difficult to reach (primary coolant system, vessel penetrations in congested areas, high radiation levels, etc.), many plants in the US and Europe have already replaced their SGs and their reactor vessel heads.

The material of choice for DMW is now Inconel 690/52/152. Inconel 690 is a nickel alloy with 58% minimum nickel, 27–31% chromium (compared to 14–17% Cr for Inconel 600), and 7–11% iron, with controlled additions of Al and Ti. Laboratory tests have shown good corrosion resistance of the material, and the plant experience with 690/52/152 in SG tubes, and buttering and cladding has been positive for over 20 years now.

The importance of surface residual stresses is also now well understood, and stress relieving heat treatments in the range of 1130 °F for several hours have been implemented.

The second damage mechanism mentioned is FAC, why is it an issue in nuclear power plants?

FAC is a classic form of metal loss, generic to all industrial piping systems. In the case of power plants (fossil or nuclear), it occurs in single-phase lines (feedwater, condensate, or steam), and in two-phase flow lines (steam and water). The event that brought special attention to FAC was a feedwater pipe rupture in 1985, which was caused by flow impingement downstream of a level-control valve, then shortly afterward, in 1986, the rupture of an 18-inch elbow to the main feedwater pump.

Which NRC documents capture the FAC occurrences?

Going back to the early FAC ruptures, the NRC had issued several documents: Information Notices 86-106 and Supplements 1, 2, and 3; 87-36, and 88-17; Bulletin 87-01; and Generic Letter 89-08 Erosion/Corrosion-Induced Pipe Wall Thinning. All plants in the US implemented FAC inspection and assessment programs, and the NRC conducted audits of several plants for FAC (NUREG 1344).

What is the current status of FAC?

Each US plant has ongoing inspection program for FAC. Interestingly, FAC is managed through EPRI, rather than ASME XI. The industry, primarily through EPRI, has developed a FAC program, where the plant systems are classified based on the potential for FAC, and individual components are identified for inspection, on a FAC risk-informed basis. The industry and EPRI continue to improve the FAC inspection and assessment program. Some example EPRI Reports on FAC include the following: State of the Fleet Project to Assess Flow-Accelerated Corrosion Program Effectiveness (2009), Effect of Pipe Size on CHECWORKS SFA Geometry Factors (2009), Statistical Methods for the Analysis of Multiple-Inspection Flow-Accelerated Corrosion Data (2009), CHECWORKSTM Steam/Feedwater (2009), Optimization of FAC Inspections (2008), Chemistry Effects on Flow-Accelerated Corrosion (2005), Recommendations for an Effective Flow-Accelerated Corrosion Program (1999), Flow-Accelerated Corrosion in Power Plants (1998), and Assessment of the Pulsed Eddy Current Technique: Detecting Flow-Accelerated Corrosion in Feedwater Piping (1997).

The third damage mechanism mentioned is neutron embrittlement. Is it the only form of embrittlement damage in nuclear power plants?

There are two types of embrittlement that could affect PWR vessel internal components. These are irradiation embrittlement, which affects the region of the reactor vessel that sees a large flux of high-energy neutrons, and thermal aging embrittlement, which may affect the cast stainless-steel parts and parts manufactured from martensitic stainless steel.

How do neutrons damage steel?

The study of radiation damage to structural materials started with the naval nuclear program in the 1950s, with much work accomplished by the then Naval Research Laboratory, the Bettis Atomic Laboratory, and national laboratories such as the Battelle Memorial Institute. Neutron irradiation is caused by neutrons escaping the reactor core and entering the thick reactor vessel wall. Above a threshold energy level, these neutrons create dislocations in the vessel steel, which eventually can result in voids, loss of toughness, and swelling. This energy threshold is encountered near the center of the reactor vessel, in a region called the vessel beltline.

The damage to materials by neutrons generated in a nuclear reactor has been correlated to the fluence (neutrons/cm^2) of neutrons with energies above 1 MeV. Such high neutron energies are limited to the center region of the reactor vessel, that is, the beltline. The effects of neutron damage were extensively studied and tested, at temperature, on reactor vessel materials. These materials, depending on the vintage of the reactor, included SA-533 (alloyed Mn-Mo-Ni steel plate), SA-508 (compatible alloyed steel forgings), SA-542 (Mo-V steel), or SA-543 (Ni-Cr-Mo steel). The microstructural damage from radiation affects primarily the material toughness and its fatigue life.

How is neutron embrittlement prevented or mitigated?

This is done on several fronts. First, the vessel design includes a neutron shield to protect the vessel wall. Second, each reactor is subject to a detailed coupon monitoring and testing program, where coupons placed around the beltline are periodically inspected and tested for evidence of embrittlement (see, for example, 10 CFR 50 Appendix H, "Reactor Vessel Material Surveillance Program Requirements"). Also, reactor vessel steel is fabricated with strict chemistry and metallurgy controls to reduce the damaging effect of neutron embrittlement and achieve the desired toughness for the reactor life (see, for example, 10 CFR 50 Appendix G "Fracture Toughness Requirements").

The operation of the reactor, its heating and cooling, is also controlled to prevent the vessel from being shocked when the metal is still cold, and more prone to brittle fracture. There is even regulation for the prevention of pressurized thermal shock (PTS), in the form of the NRC's PTS rule 10 CFR 50.61. Continuing reviews of the prevention, mitigation, and coupon monitoring programs are reported in NUREG-1511, "Reactor Pressure Vessel Status Report" and its updates.

So beyond the reactor region, the effects of radiation on materials are not a concern?

For metals this is the case because they are affected by high-energy neutrons. However, the picture is different for nonmetals. Some nonmetallic materials that are excellent gasket and packing material for so many commercial applications lose their physical and mechanical properties in a lower-radiation environment, and they are therefore prohibited in a radiation environment.

Have there been attempts at trending the causes of damage across plants?

There have been several comprehensive research projects to analyze failure data in nuclear plants, trend the data, and improve our understanding of the causes and effects of different damage mechanisms. One example is the NRC-sponsored Proactive Materials Degradation Assessment (PMDA) for light water reactor materials, the impetus being life extension. Under the PMDA program, a large body of pressure equipment and piping failure reports from several large databases were reviewed and grouped, and then analyzed using probabilistic fracture mechanics models in NUREG/CR-6936 "Probabilities of Failure and Uncertainty Estimate Information for Passive Components - A Literature Review." The causes of failures addressed were as follows:

- SSC: intergranular, transgranular, PWSCC
- Fatigue: thermal, vibration
- Flow-assisted degradation: FAC, erosion corrosion, cavitation erosion
- Local corrosion mechanisms

These studies provided statistical data for estimation of leak and rupture frequencies, the development of probability of loss of coolant accident and are captured in NUREG-1150, NUREG/CR-5750, WASH-1400, guidance for risk-based in-service inspection (Standard Review Plan Section 3.9.8), and probabilistic fracture mechanics software.

In summary, material degradation is one of the more important challenges engineers have to tackle, across industries. The nuclear power industry, like others, has had and will continue to have material challenges, but it is an industry that has approached these issues openly, sharing information across plants and internationally, at conferences, through event reports, and through NRC or industry-sponsored research reports. The industry has and continues to implement collaborative research through the NRC, national laboratories, EPRI, the IEAE, and expert consultation. In addition, each plant implements a rigorous in-service inspection (ASME XI) and in-service testing program (ASME O&M) to continuously monitor the condition of essential structures, systems, and components.

CHAPTER 7

Operability and Industry Operating Experience

Chapter 7 describes the practical challenges and solutions that plant engineers face in nuclear power plants, and the decisions made to operate the plant safely. Many of the challenges described here are not single cases; instead, they are an amalgam of actual cases experienced in plants. Chapter 7 also covers the broad experience-based knowledge, the industry operating experience (IOE) that the nuclear power industry accumulates and shares through groups such as Electric Power Research Institute (EPRI), Nuclear Energy Institute (NEI), Institute of Nuclear Power Operations (INPO), and the Nuclear Regulatory Commission (NRC). Some of the operability issues and industry experiences described here are complex and had to be resolved by an industry-wide effort. Other issues are more straightforward, but despite their simplicity we wanted to mention them because they also constitute the day-to-day challenges encountered by nuclear power plant engineers, and they have to be solved with the same level of thoroughness as the more complex issues.

7.1 OPERABILITY

What is meant by operability?

In all industries, engineers are called upon to determine whether a structure, system, or component (SSC) that has undergone some damage, or has been subjected to an abnormal condition, is fit for continued safe and reliable service, or whether it should be placed out of service. In the chemical process and hydrocarbon industries, this evaluation process is called fitness-for-service (FFS), or fitness-for-purpose. In the nuclear power industry, it is called operability assessment or operability determination and refers to the assessment of degraded or nonconforming conditions.

What rules govern the determination of operability of an SSC?

There are two sets of rules: those of the NRC, and those of ASME XI and the ASME Operations and Maintenance (O&M) code. The NRC guidance regarding operability determination was outlined in Generic

Nuclear Power Plant Safety and Mechanical Integrity
ISBN 978-0-12-417248-7

Letter 91-18. This guidance has since been folded into the NRC Inspection Manual Part 9900, and plants were informed of this change through a regulatory issue summary 2005-20. From there, and with guidance from NEI, plants have developed plant-specific operability determination processes. This process is closely linked to 10 CFR 50.59 "changes, tests, and experiments."

What are the main features of the NRC's inspection manual part 9900?

Its scope is made clear from the get-go through its title: "Operability Determinations and Functional Assessments for Resolution of Degraded or Nonconforming Conditions Adverse to Quality or Safety." It is meant to be internal guidance to the NRC inspectors, but of course is quite useful to any engineer in performing this type of assessment.

The guidance applies to SSCs that have a technical specification (TS) function or those that provide support functions to the TS SSCs. The expectation for an operability determination is to establish a reasonable expectation that the SSC is operable, where reasonable expectation corresponds to a high degree of confidence of good performance based on solid facts and a sound logic.

How much time is available between the time the degraded condition is identified and the time it is resolved?

An immediate determination is needed to decide whether there is a reasonable expectation of operability. If this judgment cannot be made without additional study, then the SSC must be declared inoperable. The second stage is referred to as prompt determination of operability during which a more detailed evaluation can be conducted. There are no explicit time limits, but in the case of highly significant safety-related SSCs, the immediate determination can be expected to be done in a matter of hours, and the prompt determination "can often be done within 24 hours of discovery even if complete information is not available. If more time is needed to gather additional information (such as a vendor analyses or calculations) the licensee can evaluate the risk importance of the additional information to decide whether to prolong the operability determination. TSs completion time is one factor that can be used in determining an appropriate time frame within which a prompt determination should be completed" (source: NRC Inspection Manual Part 9900).

Does the operability determination have to meet the current licensing basis?

In some cases described in Part 9900, more lenient criteria can be applied on an interim basis until a permanent repair is put in place. Examples of more lenient, yet still safe, criteria are the use of Appendix F to quantify margins to failure or the crack-like flaw assessment criteria of GL 90-05 or the use of ASME XI Code Case N-513 under the conditions specified in RG 1.147.

Who makes the call of operability or, to the contrary, of inoperability?

The NRC's Part 9900 gives this responsibility to a senior licensed operator on the shift crew.

Are these operability challenges unique to nuclear power plants?

Generally, these operability challenges are not much different from those faced by engineers in other industries, where their assessment is referred to as FFS. There are unique aspects to the determination of operability in a nuclear power plant environment:

- The regulatory requirement for an immediate determination, with its expectation of an hours-short time frame, and the follow-up prompt determination, which, if they cannot be met, will cause the SSC to be declared inoperable; this could in turn cause a plant shutdown.
- The level of documentation of the operability determination, which must establish a high degree of confidence of operability. In nuclear power plants, the challenges must be thoroughly described, documented, analyzed, and resolved. Formal condition reports are documented in each case.
- The operability determination must address not only normal operation but also operability under hypothetical design accidents that are unique to nuclear power plants such as large design earthquake (the safe shutdown earthquake), transients from postulated pipe breaks, accidental large ambient temperatures, etc.
- A unique aspect of nuclear power engineering is what we can be called "what if." If a condition outside the design basis is identified, say a failed seismic brace, then the engineer must analyze what would have happened if a postulated accident (in this case a design-basis earthquake) would have occurred with the failed hanger, even though the accident did not occur, and the condition was repaired (in our example, the hanger was replaced).

In this chapter we describe a series of challenges and their resolution. We will arrange them by subsection on the basis of the class of SSC affected: piping and tubing systems, raceways (conduit and cable trays) and ducts, mechanical equipment, and structures.

7.2 PIPING AND TUBING SYSTEMS

Challenge 1: Moisture Damage

The ingress of moisture around outdoor penetrations in a pump room had damaged the coating of a pump discharge pipe. The remaining piping, away from the penetrations, was in good condition. Engineering was requested to evaluate the integrity of these affected sections of the piping system and their supports.

Solution:

Upon identifying the damaged coating, the pipe surface in the affected area was prepared for ultrasonic inspection, and the pipe wall thickness was measured over a large area around the damaged coating. The wall thickness turned out to be equal to and in some places greater than the nominal wall. It is not unusual to measure a wall thickness greater than the nominal, given the mill tolerance on the pipe wall, which is stated in the material specification. This is even more common in fittings. The condition was therefore found consistent with the design basis and acceptable. The coating was reapplied and inspected. The penetration seal was repaired to prevent further moisture, and the pipe continued satisfactory service.

Challenge 2: Inspections Following a Water Hammer

While inspecting pipe supports following a water hammer, two support baseplates were found to be partially pulled away from the concrete wall. Some concrete spalling was found around the anchor bolts. In both cases, the gap between the baseplate and the wall exceeded the permitted design gap.

Solution:

Structural damage to pipe supports is a typical sign of large water hammers. Usually the welded pipe itself fares well, owing to its high ductility, and it is the more brittle components in the load path, in this case the concrete anchor bolts, that suffer. The rest of the line was inspected to determine the extent of condition. The engineer must always search for the extent of damage, beyond the initial indications.

The pipe was lifted using a power jack bearing against a temporary plate on the floor, with the pipe-to-jack contact area protected with a pad to prevent bearing damage. This relieved the load on the baseplate and permitted a detailed inspection of the baseplate, the wall, and the anchor bolts.

The observed damage was used to back-calculate the magnitude of load experienced during the water hammer, and the back-calculated load was found to be consistent with the engineering prediction of the transient. The baseplate was repaired before the system was put back into service. The cause of the water hammer was determined, and procedures were modified to prevent its recurrence.

Challenge 3: Lesson from Flow-Induced Crack in Socket Weld

A pipe sprung a leak at the pipe-to-valve socket weld in the form of a fine mist of pressurized water.

Solution:

The cause of the leak was traced to a high-cycle fatigue crack at the toe of the socket weld, which was caused by flow-induced vibration (FIV). The piping and the valve configuration were modified to prevent the vibration. In addition, the new socket welds were made concave to prevent stress concentrations at the toe of the weld, and the socket weld profile was changed to a 2 to 1 taper. This last precaution results from what have been learned through EPRI-sponsored vibration testing: A socket weld twice as high (pipe-side) as it is wide (socket-side) has a significantly improved fatigue life. Once the installation was completed, the new valve was monitored for vibration, in accordance with plant procedures that are based on the ASME O&M Part 3 criteria.

Challenge 4: Run-or-Repair Decision for Corrosion

Localized wall thinning corrosion was detected in a 10-inch sch. 40 service waterline as part of the plant in-service inspection program. The 0.365-inch (9 mm) nominal wall had been reduced by nearly 30% over a narrow area, near a weld. UT readings confirmed that the pipe wall surrounding the thin area was near-nominal, within the 12.5% mill tolerance for this pipe. This particular corroded area had been monitored for several years, and the corrosion rate was established based on successive UT wall thickness measurements.

Solution:

In case of wall thinning corrosion, the plant has a formal procedure to evaluate the inspection results and determine the remaining life of the corroded component. The method of evaluation of corrosion depends on the type of corrosion damage. If the damage is by wall thinning (corrosion or erosion), the plant will typically evaluate the thickness readings based on ASME XI Code Cases (CC) N-597 or N-513. It is important to keep in mind two important conditions for applying ASME XI CC: First, the CC must be approved for use at the plant. Second, the CC must be applied within the limitations of R.G.1.147. There are many CC not approved by the NRC; these are listed in R.G. 1.193. In the case presented here, a remaining life was established based on a conservative corrosion rate and considering all the normal operating and postulated accident loads. The joint was then cut out and was replaced at the next refueling outage, well ahead of its predicted end of life.

Challenge 5: Condenser Wall Vibration

A vacuum sensing instrument tubing connected to a plant main condenser wall sheared off, causing a loss of vacuum trip of the main turbine generator, resulting in turn in a reactor trip.

Solution:

First, notice the interlocking of systems in a nuclear power plant: When an abnormal situation takes place, in this case loss of vacuum in a sensing line, automatic protections come into play, up to and including tripping the reactor if necessary. The rupture here was traced to vibration of the condenser shell wall when operating under negative pressure. It was concluded that the instrument line was not routed in a manner to accommodate the vibration of the condenser shell. The routing of the vacuum sensing line outside the condenser wall was modified to adequately account for the shell movement. Three connections for instrument sensing lines were provided at each location, for redundancy. Other tubing and piping connections were inspected for damage indications and vulnerabilities to static movements as the shell pulls inward under negative pressure, and to vibration during operation.

Challenge 6: An Out-of-Tolerance Snubber

During periodic in-service testing, a snubber was found to have too high a drag force outside its drag force tolerance.

Solution:

Since the snubber was reported to have high drag, it would resist the piping thermal movements in the restraint direction. The stress analysis of the line with the high-drag snubber was rerun conservatively assuming the snubber was acting as a rigid restraint. It was found that despite this conservative assumption no significant thermal load was generated, even if the snubber had fully locked. Since it was not needed, the snubber was replaced by a rigid strut, the design stress analysis was revised to reflect the change, and the system was returned to service.

Challenge 7: Temporary Removal of a Hanger

To allow access to inspect a pipe lug weld, a spring hanger had to be removed temporarily. The spring would be replaced in position after the inspection.

Solution:

In this case, an Engineering evaluation was performed to determine if a temporary support was needed to support the line and maintain its slope while the permanent support was removed for inspection. It was found that the load would be transferred to the nearby supports on each side, and that these adjacent supports had sufficient capacity to support the line and maintain its slope. Note that because it is a spring hanger that had to be removed, its temporary removal will not affect the system qualification for postulated seismic or other dynamic loads. These dynamic loads had to be assumed to happen during the short period of time when the hanger is removed. The support was reinstalled to the original design.

Challenge 8: Consequence of a Pressure Uprate

Following the replacement of steam generators (SG) in a pressurized water reactor (PWR), the design pressure of the main steam line was increased from 1000 to 1007 psi. Engineering had to evaluate the effect of this 0.7% increase in pressure on the design of the steam lines.

Solution:

First, we review of the pressure rating of the system to determine whether it can sustain the 7-psi increase. Next is the question of whether the initial hydrotest would still be valid despite the new increased pressure. Then, with a change in pressure comes a change in flow rate, which is investigated for the potential for acoustic resonance. Another aspect to investigate is the postulated pipe whip impact force on the main steam whip restraint, which is a function of the line pressure. It turned out that this whip effect was the controlling design margin, but it was acceptable. The plant specifications, calculations, and design documents were then revised to reflect the change of 7 psi.

Challenge 9: Trapped Air Pockets

This challenge was a generic concern, which applied to all power plants, and was formally raised in 2008, through NRC Generic Letter GL-2008-01. In simple terms, the question asked was whether air pockets trapped at high points in water lines could be entrained, causing pressure imbalance when swept into downstream pumps, cavitation issues to the suction side of pumps, or pressure transients (water hammer) mainly on the discharge side.

Solution:

First, the industry reviewed the systems and locations where such air entrapment could take place. The emergency core cooling system and the containment spray (CS) were two systems reviewed. Tests and numerical simulations were conducted to better understand the pressure oscillations caused by compressed air pockets in water-filled lines. Then, nondestructive examinations were conducted to identify the actual location and size of air pockets. Following inspections, trapped air pockets were either vented, in some cases adding high point vents, or analyzed and qualified.

Challenge 10: Overpressure is within Allowance

During restoration of a system in preparation for reactor start-up, a piping became overpressurized for a short duration. This condition was recognized in the control room and promptly resolved in a matter of minutes.

Solution:

Review of pressure records indicated that the maximum overpressure experienced in the system was less than 10% over the design pressure. This overpressure and the short time it lasted were within the overpressure allowance in the design code, in this case ASME B31.1. In addition to meeting the code, the line was inspected for any evidence of damage before it was returned to service.

Challenge 11: Temporary Changes of Configurations

Engineering was asked to evaluate the temporary removal of a vertical brace to replace a valve gasket.

Solution:

The removal of the vertical strut would cause an increase in load on the adjacent spring hanger. This challenge was resolved by pinning the spring, after it was verified that the spring and the backup structure could sustain the additional load, within their design limits. Second, the removal of the vertical brace would change the seismic and pipe break design basis of the line. The system was therefore placed in the "inoperable" status, until the valve gasket was replaced, the brace was reinstalled, and the pin removed, at which point the system was returned to service.

Challenge 12: Post-Water Hammer Actions

During a surveillance test, a valve was opened with tank deadhead on its upstream side and negative pressure on its downstream (main condenser) side. A water hammer (loud noise, pipe movement) was reported by the test personnel.

Solution:

The valve alignment sequence at the onset of the water hammer was traced on the system P&ID and isometrics to understand the physics of the event: Was it a liquid system water hammer, a cavitation water hammer, or a slug flow? This study confirmed that, indeed, a cavitation water hammer would be expected to occur under these conditions, if the valve was opened too quickly. The line was analyzed to simulate the event. In parallel, the piping system and its supports were closely inspected visually. Although the calculated stresses were well within the design code limits and the visual inspection showed no sign of damage, points of high stress were volumetrically inspected more closely for evidence of damage. The line was pressure tested and was then returned into service. Operating procedures were revised, with the corresponding operator training, to prevent recurrence.

Challenge 13: FIV ... Again

A pipe U-bolt downstream of a pump discharge nozzle was found to be broken.

Solution:

The failure was traced to a change in pump operating speed, which changed the pressure pulsation frequency in the line. This change of frequency of the pressure pulse caused a new structural resonance, which eventually resulted in the fatigue failure of the U-bolt.

Once the cause was understood, parallel systems were reviewed and their supports closely inspected. The support arrangement of the affected line was modified to eliminate the structural resonance. During restart and for part of its operation, the system was monitored for vibration, following the methods and criteria of ASME O&M Part 3. The monitoring confirmed that the design modification of the support system had resolved the issue.

Challenge 14: Braided Hose Replacement

A section of braided metallic hose was visibly vibrating in service. There was a concern that this vibration would eventually fail the hose connection to the header pipe.

Solution:

First, it was determined that the observed vibration of the braided hose was caused by a structural resonance between the hose and the driving hydraulic frequency. To detune this resonance, the braided hose was replaced with lighter tubing, as shown in Figure 7.1. The tubing natural frequencies were verified to be different from the hydraulic driving frequency, and a hammer bump test on the tubing was conducted to confirm the calculation results. The bump test was done by striking the tubing with an instrumented hammer of the right weight and tip rigidity. Then, the vibration acceleration was recorded and analyzed through a fast Fourier transform to determine the natural beam frequencies of the tubing system. The tubing also had to be routed and supported to accommodate the large thermal expansion movements of the vertical header pipe to which it is attached. This was achieved and verified by analysis, comparing the expansion stresses with the applicable ASME code allowable stresses.

Challenge 15: Abandoned-in-Place Support

An out-of-service 1-inch piping section located over a piece of equipment was removed, but the pipe supports were abandoned in place after the pipe

Figure 7.1 Tubing replaced a flex hose to prevent vibration-induced failure.

was removed. Shortly afterward, a piece of angle from one of the abandoned pipe supports fractured and fell on the equipment and associated tubing causing damage and leading to a plant trip.

Solution:

By inspection of the broken support angle, it was established that the failure was caused by vibration-induced fatigue. This fatigue may have started when the pipe was still installed in position, but it became larger when the support was no longer connected to the piping system and was instead hanging out as a cantilevered beam. This type of failure leads to a broader lesson learned, which is to take into consideration the structural ramifications of modifying a design, even when it is by abandoning something in-place.

Challenge 16: Nonmetallic Joint

A CPVC-to-steel threaded joint leaked upon system heat-up. While this was a nonsafety system, it was still important to understand the cause of this leak to prevent its recurrence.

Solution:

The first step was to inspect the joint to make sure that it was properly made, following written procedures, by qualified personnel. This was confirmed to be the case; the workmanship was not in question. After several causes of leakage were considered, analyzed, and discarded, it was determined that the leak was caused by the differential thermal expansion between the CPVC female fitting and the much lower expansion of the steel male insert. The joint was replaced. Other systems were reviewed for similar plastic-to-metal joints, but it was determined that this was a unique case due to the larger operating temperature, the other joints being operated at atmospheric temperature.

Challenge 17: Leaking Expansion Joint

Two parallel pumps take their suction from a vessel. The vessel-to-pumps suction pipe is 30 inch in diameter and consists of a vertical drop from the vessel bottom nozzle, through a stainless-steel (SS) expansion joint, and then to a header feeding each of the two pumps. After several years of service, the expansion joint developed a dripping leak from a tight crack. The pumps and piping are nonsafety related.

Solution:

As an immediate action, the pipe was inspected, all around the leaking crack, to identify any other damage, either cracking or wall thinning. None was found. The root cause of the crack was traced to a defective pipe support, which caused the movements at the expansion joint to be different than what it was designed for. Having verified the extent of condition and understood its cause, mitigation measures were put in place until the joint could be replaced at the next outage. A shield made of welded steel plates was designed and bolted in position around the leaking expansion joint (Figure 7.2). The shield had to meet several design requirements:

(1) It should act as a leak box, encapsulating the leaking crack.

(2) It should not affect the system operating parameters (flow, temperature, thermal expansion/contraction, etc.).

(3) It should be of a shape and weight that is qualified under the design code, in this case ASME B31.1, as well as qualify the reactions on the adjacent nozzles and the joint itself.

(4) It should be sufficiently large and strong to act as a jet shield, should the crack extend beyond its current condition.

(5) It should allow for detection of an increase in leakage flow, should it happen.

(6) The materials used to fabricate the shield and the welds should have a design life well beyond the required operating life, even when wetted by the leakage drip.

(7) Nonmetallic sealant injected into the leak box portion of the shield should be compatible with the environmental conditions and the expected design life of the shield.

(8) Despite all these measures, the hypothetical case of a full guillotine rupture of the joint should be analyzed and remedial actions put in place, for the remote possibility of such an occurrence.

Figure 7.2 Leaking expansion jet shield and leak box.

Challenge 18: Pipe Uplift

A 16-inch run of pipe operating at high temperature, near 600 °F, and subject to slow insurges and outsurges of hotter water at low flow rate, slowly bowed up during system start-up, lifting off its support.

Solution:

The vertical uplift of the pipe was traced to stratified flow in the pipe. During start-up, the ends of the pipe are exposed to two reservoirs with a temperature difference of as much as 300 °F. As the water surges from one end or the other into the pipe, the difference in pressurized water density caused by the 300 °F difference in temperature forced the flow to stratify, with the hot-flow plume slowly flowing above the colder water. This top-to-bottom temperature gradient caused the pipe to uplift. The solution to this condition was to modify the heat-up procedure to reduce the gradient, but also to analyze the stress effects that such a gradient and uplift would have on the pipe. Interestingly, a similar uplift effect was reported in a horizontal steam pipe with stagnant condensate at the bottom. As steam is introduced in the system, it flows on top of the colder condensate, causing a top-to-bottom temperature gradient, which uplifts the pipe.

Challenge 19: Localized Erosion

A digital radiography of butt welds identified several welds, which had wall thinning right at the downstream side of the butt weld.

Solution:

A review of these weld construction records identified that at construction these welds had met the code. A close look at the in-service inspection radiographs pointed out weld protrusions on the inner diameter. Because this line had an unusually large flow rate, the protruding welds caused local vortices on the downstream side, leading to local erosion after many years of service. All welds in similar service were inspected, and welds with wall thinning or inner protrusion were cut out and replaced.

Challenge 20: Generic Issue, the Corrosion of Buried Pipe

Engineering was asked to develop a buried inspection program to monitor the corrosion condition of safety-related buried pipes in the plant.

Solution:

This is a generic issue, common to all nuclear power plants: As the systems age, after over 30 years or more of service, the corrosion condition of underground piping needs to be measured and remediated where necessary. As is the case with many generic issues in the US nuclear power industry, the effort was spearheaded by EPRI, NEI, and INPO, with close oversight by the NRC. A Buried Pipe Integrity Group (BPIG) was formed by EPRI. The BPIG grouped US and several international utilities to develop a correct, complete, and consistent approach to tackle the challenge. In addition, NEI and INPO developed expectations and guidance to help in the implementation of the inspections. The BPIG developed criteria to select which lines should be inspected, where, when, and how. An ASME XI Code Case N-806 was written to help utilities evaluate the wall thickness readings that came out of the inspections. A new ASME XI Section IWA-5244 Buried Component Leakage Test was developed, and a new Code Case N-776 was written to address ground surface examination for leaks. At this point, with inspection methods and criteria in place, the inspection programs started, and the BPIG, NEI (Guideline NEI 09-14), and INPO helped in sharing lessons learned among plants. The NRC was involved throughout in reviewing the adequacy of the buried pipe inspections and assessments. This includes the NRC Chairman tasking the NRC staff (Task 09-146), periodic NRC staff reviews, meetings with industry, and plant-specific and generic review reports.

7.3 RACEWAYS AND DUCTS

Challenge 1: Routine Maintenance Inspection

Class 1E cables are routinely inspected for general condition and conformance to design drawings. In a few cases, the tie-wraps that secure cables together were in need of replacement.

Solution:

An engineering evaluation was performed to determine the cause of the damage and take remedial action to prevent recurrence. The missing cable tie-warps were replaced with better heat- and radiation-resistant material tie-wraps.

Challenge 2: Installation of New Conduit

Maintenance is to install new runs of conduits and supports, and to rework multiple existing conduits as well as junction boxes.

Solution:

A design change is issued to update the conduit drawings and indicate the location and details of the new conduit supports. Electrical and civil-structural calculations are developed to qualify the new conduit system and its supports to the plant standards, and to verify that the conduit is not the source or the target of seismic interactions. The modification package undergoes a multidiscipline review, and the design change is planned, implemented, and as-built verified.

Challenge 3: Commodities Clearance Inspections

A condition existed where a conduit was found to be too close to pipe insulation.

Solution:

Although this condition was determined to have no adverse effect on the pipe or the conduit, the conduit was rerouted to maintain the commodities clearance requirements. The commodities clearance requirements are imposed to prevent interference from thermal expansion and rattle impact in case of a seismic event, among other reasons.

Challenge 4: Good Housekeeping

A tie rod in a run of HVAC duct was found to be deformed.

Solution:

This may have been a damage dating back from construction that was discovered later during periodic inspections. The as-is condition was analyzed and the damage to the tie rod was found to be acceptable. The tie rod was replaced.

Challenge 5: Broken Support Stud

A routine walkdown of an HVAC system identified a broken stud on a duct support.

Solution:

It was apparent by inspection that the stud failed by atmospheric corrosion. Further inspection identified signs of humidity on the wall. Actions were implemented to remove the humidity from the area. The stud was replaced, and all other studs and bolts in the room were inspected; they were found to be acceptable. The room penetration seals were inspected and repaired as needed, and now the area is routinely checked for signs of humidity or rust.

Challenge 6: Rerouted HVAC Duct

Several runs of HVAC duct had to be rerouted to accommodate the removal and replacement of a large vessel.

Solution:

A new route was identified for the ducts. The system flow and pressure drop calculations were verified for the new layout. Once the flow and pressure design were confirmed, the new duct route was analyzed for pressure differential, weight, and seismic loads. The sheet metal duct and stiffeners were designed in accordance with ASME AG-1. The layout was verified to be free of adverse seismic interactions. Following installation, the as-built duct was verified, and the duct was leak-tightness checked prior to be put into service.

Challenge 7: Aged Cable Wraps

During a housekeeping walkdown of a nonsafety building, white cable wraps (cable ties) tying a bundle of cable in a cable tray were found broken.

Solution:

As a result, all cable ties in the building were inspected, and a handful of others were found to be broken. An investigation of the material revealed that the tie wraps had become brittle with time and eventually broke. As a result, the cable ties in this building were replaced with a new material better resistant to the elements and with the correct temperature and flammability rating. Their new color, blue, permits the plant staff to recognize these new ties and over time replace all the old white ties with the new, blue ones—another example of the proactive approach common in nuclear power plants.

Challenge 8: Incomplete Manufacturer Drawing

A new pump was received on site, but the pump assembly drawings did not indicate the supports of the new electrical conduit.

Solution:

Prior to installing the pump, engineering laid out, analyzed, and qualified the conduit supports, and developed a design change to the pump manufacturer drawing to indicate the new pump conduit supports, and an as-built was conducted to reflect field installation of the supports.

Challenge 9: Updating Old Criteria to Keep Up with the Times

A new generation of IT communication cables slightly exceeds the old plant linear weight limit on cables.

Solution:

Engineering retrieved the technical basis of the old cable linear weight limit and found that it included large margins, of 10 or more, over what is needed to qualify the cable runs and supports. The increase in weight of the new IT supports is well below this margin; this was documented in new design calculations and approved for installation.

Challenge 10: Improved Temporary Cooling Unit

As part of the plant continuing improvement effort, a new air handling unit (AHU) was identified that can provide cooling during outages without the need for multiple temporary coolers, temporary flexible ducts, and temporary installation scaffold that were needed for the existing AHU.

Solution:

To install the new AHU, a flexible connection is needed from the existing damper on top of the AHU to connect to a containment recirculation duct directly above the AHU. Engineering designed a flexible connection from the new AHU to the existing ductwork. The new flexible duct was sized for flow and pressure, and qualified for its environment, and the seismic loads and movements. Drawings were revised to incorporate the field changes.

Challenge 11: Damp Vault Wall

Inspections of cable supports in a corner of a vault identified rust on the concrete anchor bolts.

Solution:

A damp corner wall was identified in the vault and eliminated by draining and sealing. The bolts were replaced, and the room was added to the standard rounds and visual inspections to make sure that the condition will not repeat itself.

Challenge 12: Replaced Bolts

After multiple untorquing and retorquing of anchor bolts retaining often-used manhole covers, Maintenance approached Engineering to replace the bolts.

Solution:

Engineering concurred that repeated untorquing and retorquing of anchor bolts can lead to stripped threads and bolt hole damage. In this case of manhole covers, the bolts were replaced with bolts of the same size but of a higher standard. The old bolt locations could not be changed due to configuration limitations and the bolt holes had to be repaired. The concrete bolt hole repair plan was field tested and found to be acceptable for repeated untorquing and retorquing of concrete anchor bolts of the type used.

Challenge 13: Baseplate Bearing Surface

A conduit baseplate support had less than the minimum 80% bearing surface contact required by the plant specification.

Solution:

The conduit was found to be a spare, carrying a single cable, and the walkdown inspection showed that the spare conduit support in question is attached to a concrete column with two steel baseplates. The second baseplate had adequate bearing surface and, by itself, was sufficient to accommodate all the design loads imparted by the conduit. In any event, a work order was created and the support baseplate was grouted to provide an acceptable bearing surface.

Challenge 14: Control of Torque

In nuclear power plants, torquing of safety-related mounting hardware for electrical components, even of screws, is performed under controlled procedures. Maintenance had requested Engineering to provide a torque tolerance for tie-downs of electrical wiring, components, and mounting hardware.

Solution:

Engineering worked with the manufacturer and acquired the data on the proposed torque tolerance, which was then made part of the maintenance procedures and maintenance training.

7.4 MECHANICAL EQUIPMENT

Challenge 1: Out-of-Calibration Torque Wrench

As part of the plant's periodic calibration testing, a torque wrench was found to be out of calibration.

Solution:

The log of work orders performed with this torque wrench since it was last used within calibration was retrieved. In parallel, it was determined that the out of calibration would tend toward overtorque. The joints that were overtorqued with the wrench in question were analyzed. The torque was converted to a preload on the bolt, and the preload was analyzed for stress in the bolt and leak tightness of the joint. The results of the analysis were well within acceptable limits with the overtorque. The bolts were accepted as is. The torque wrench was recalibrated.

Challenge 2: Overtorqued Bolt

According to plant procedures, the maximum torque for the size and grade of a certain hex bolt is 110 $\pm$ 5 ft-lb. A condition was identified where a pipe support drawing called out 130 ft-lb instead of 110 ft-lb.

Solution:

This discrepancy between the general plant procedure torque of 110 ft-lb and the as-installed pipe support torque of 130 ft-lb was evaluated. It was confirmed that the field-torqued 130 ft-lb was indeed in accordance with the torque specified in the pipe support drawing. It was also determined that the 130 ft-lb torque was acceptable, provided the bolt is not reused if it was to be removed in the future.

Challenge 3: Double-Seal Design

During periodic testing, the inner O-ring seal on a pressure vessel was found to have a leak.

Solution:

This pressure vessel flange sealing system was designed to have two seals, one inner and one outer, and a leak detection capability to detect a leak in the annular space between the seals. This design would allow the inner seal to leak and have the outer seal contain the leak, with alarmed leak detection. A troubleshooting plan was prepared and implemented to confirm and pinpoint the inner seal leakage and to confirm the integrity of the outer seal and the leak detection system. While the inner seal leakage was well contained by the outer seal, the condition was repaired.

Challenge 4: Cracked Compressor Housing

During periodic inspection, a surface crack was identified in a safety-related compressor motor housing made of cast iron. Engineering was requested to identify the cause of the crack and develop a repair plan and actions to prevent recurrence.

Solution:

The cause of cracking was determined to be fatigue caused by pressure pulsing inside the compressor. No replacement could be found within a reasonable time for the compressor motor housing. Engineering designed a steel collar plate, which provided three benefits: (1) stiffen the housing to reduce the vibration amplitude, (2) reinforce the housing, relieving the load path away from the cast iron, and (3) the collar plate contained a perimeter gasket seal to act as a leak-containing box, should the crack continue to propagate. The modification was tested and found to be acceptable to contain the leak and maintain the pressure integrity. A permanent replacement was planned for implementation at the upcoming outage.

Challenge 5: Leaking Threads

The visual inspection of a threaded nozzle on a hot vessel indicated deposits of chemicals on the outer surface of the threads.

Solution:

The system was shut down and cooled, and the insulation was removed from the vessel head for a thorough inspection of the nozzle and the vessel head around the nozzle. An air leak test with bubble solution confirmed the leak at the threaded nozzle. The threaded nozzle was seal welded with a fillet weld. The weld was performed in accordance with plant procedures for vessel repairs, including compliance to ASME III for design, ASME IX for welding, and ASME XI for repair procedures. The repair was leak checked before the insulation was repositioned, and the system returned to service. Engineering was made aware of standard ASME PCC-2 "Repair of Pressure Equipment and Piping," which, while not a nuclear standard, is a nice compilation of repair options.

Challenge 6: Leaking Gasket

Boron residue was found on the carbon steel bolts of a flange in the CS heat exchanger. This is evidence of a leak of hot borated water. As the water containing boric acid leaks out of the system, it flashes to steam and leaves behind the boron residue.

Solution:

Engineering determined that there is a risk of corrosion of the carbon steel bolts by boric acid. The bolts were removed and inspected for corrosion in the bolt body and threads. Surface rust was cleaned off, and the corroded bolts were replaced. Later, the bolts were replaced with SS material.

Challenge 7: Sheared Bolt

Leakage from a valve was initially identified as a flange leak. A closer inspection revealed that the leakage was actually from the valve shaft. An attempt was made to tighten the bolts on the shaft cover to stop the leak. When the wrench was applied to tighten the first bolt, the bolt sheared and the leak increased.

Solution:

A clamp was installed to retain the shaft cover and stop the leak temporarily. This was a compensatory measure to permit continued operation safely. Later on, when the conditions were right, the system was isolated and depressurized, and a leak box was installed around the valve and piped to a drain tank. The broken bolt was replaced, and the other bolts holding the valve shaft cover were inspected. Since all bolts showed some sign of damage, they were all replaced. The valve was leak tested with the new bolts in place. The repaired valve was then mounted onto the system and the system was leak checked before return into service.

Challenge 8: Cracked Expansion Joints

As part of the installation of new low-pressure turbines, the existing steam inlet expansion joints were inspected. The inspection revealed that three of the four joints had crack-like indications on the surface.

Solution:

The cracks were inspected by surface and volumetric examination. The inspections confirmed that the cracks were shallow fatigue cracks due to cyclic thermal expansion–contraction. The piping system was analyzed, and differential movements were calculated at the ends of the expansion joints. The end movements, number of cycles, and operating conditions (pressure, temperature, fluid, and environment) were used by the expansion joint manufacturer to select better-suited joints. The selected joints were then entered into the piping stress model, the system with the new joints was qualified, and the final analysis was documented. The expansion joints were then replaced following the manufacturer procedure by qualified technicians to prevent damage or distortion during installation. The new joints were leak checked in position before returning the system to operation.

Challenge 9: Expansion Joint Improvement

A "what if" analysis indicated that the failure of a 30-inch expansion joint in the turbine steam system would cause a change in pressure drop and flow rate across a low-pressure turbine that may damage its blades. This would be a crippling failure that would take time to repair.

Solution:

In order to prevent the failure of the expansion joint, Engineering decided to replace the existing bellows liner with a thicker liner, with a tapered shape to improve flow and minimize turbulence. The material was verified to be compatible with the process chemistry, pressure, and temperature. The joint manufacturer concurred with the change and provided new design drawings for this design change.

Challenge 10: Improved Tank Fill Time

Engineering was asked to develop a more effective means to fill the plant safety-related accumulator tanks, an operation that is performed at the end of each outage. The intent was to expedite completion of this activity in a more timely and safe manner, and take this activity off the outage critical path.

Solution:

Several options were investigated and the optimum solution agreed upon was to develop an alternate fill path with its own safety-related pump. A new pipe route was selected. The system flow analysis was performed to hydraulically size piping, pumps, and valves, accounting for frictional losses. It was calculated that the tank could be filled in 4 h, which was well below the current fill time. Once the new system was installed, it was tested and confirmed that the tank fill time was indeed within the calculated predictions.

Challenge 11: Modified Flow Path

The installation of a new, upgraded low-pressure turbine called for changes to the condensate flow to increase flow of the heater drain pump.

Solution:

The flow of the heater drain pump was increased by modifying the pump discharge head, using the condensate recirculation valve in a new alignment. The valve internals and the actuator were replaced to reduce potential cavitation that may result from the increased flow rate. The new pipe route and its supports were analyzed and qualified to ASME B31.1. The system was checked for potential acoustic resonance introduced by the new, higher flow rate. Because FIV is difficult to accurately predict in all cases, the system was also monitored for vibration during start-up.

Challenge 12: Steam Impingement Plate

A large steam pipe discharges into the main condenser. The condenser is protected from steam erosion by an impingement plate that is reaching its end of eroded life. Engineering was to provide a design to replace the impingement plate and propose improvements to the current design.

Solution:

Engineering first evaluated the impingement angle; the evaluation concluded that the current angle is nearly optimum. Second, Engineering checked the impingement plate material. Several materials were investigated, and for this application it was determined on the basis of cost–benefit that an ASTM A283 Grade C carbon steel plate would be the best, and the plate was replaced.

Challenge 13: Valve Cavitation

A flow control valve positioner was damaged as a result of FIV over a long period of time. As a result, the valve would not fully close.

Solution:

Engineering determined by analysis and observation that the flow through the valve was choked and was causing cavitation when the valve opening was within a certain range. For short-term operation, Engineering recommended that the valve opening be limited to the range that prevents cavitation. At the next outage the valve actuator was replaced providing a permanent solution. In addition, clamps were installed on the outlet tubing to assist in stiffening the tubing line and dampening the vibration.

Challenge 14: Early Warning Initiative

As a result of continuing vibration problems on several steam systems, Operations requested the installation of wireless vibration equipment to monitor various plant SSCs and provide early warning of vibration.

Solution:

Engineering implemented a wireless vibration monitoring system on the selected SSCs. The key questions that guided the effort are described here.

- What system to select and what to monitor? It was decided to monitor and record accelerations; these could be transformed into vibration velocity or amplitude by signal integration. Another valuable decision was to also monitor pressure, so as to correlate changes in vibration signal with pressure pulsing or pressure transients that occurred at that same time. Typical installation details for battery-powered sensor hubs were provided. The hubs provide power to accelerometers that are mounted on SSCs to monitor vibration. Wireless monitoring were considered and utilized based on current technological capabilities.
- Which locations to monitor? The sensor locations that require attachment details to permanent plant structures were identified by collaborative input from Operations, Maintenance, and Engineering. Nearly 100 monitoring locations were identified. New supporting structures for these hubs were designed and installed. Pumps were already monitored for vibration for early warning signs of bearing degradation or shaft alignment issues.
- Which criteria to use in analyzing the data? When recording data, it is necessary to have, ahead of time, screening criteria to decide what level of vibration is acceptable and what is not. Once the data collection

started, there were periodic reviews to study the results and decide how best to adjust monitoring.

Challenge 15: Pressure Uprate

A safety-related system has to be uprated from 1750 psi (current set point on the pressure relief valve) to 1780 psi: a pressure increase of 1.7%.

Solution:

What at first view seems like a small pressure increase cannot be taken lightly in a nuclear power plant environment, where the consequences of every change must be thoroughly investigated. First, it was decided that the new design pressure would be set at 1960 psi, approximately 10% over the maximum operating pressure of 1780 psi. Rules for re-rating are provided in ASME XI IWA-4330. Engineering had to first review the effect of this change on the thermohydraulics of the system (flow rates, flow regime, pressure drops, temperatures, etc.). Then, Engineering had to uprate the piping system and components. Stress analyses of record were revised for the new design pressure. The relief valve set points was changed to 1960 psig by replacing the valve springs and washers. Engineering evaluated the effect of the increased discharge flow rate on the downstream tanks. Once all the engineering studies were completed and the change was approved, the system was pressure tested at a pressure based on the new design pressure, in accordance with IWA-5000. The ASME certificates were updated as required in ASME XI. Finally, the system was monitored during start-up to confirm the hydraulic parameters and check for evidence of vibration.

Challenge 16: Misaligned Bolt Holes

The holes for bolts on a new pump baseplate do not line up with the existing concrete anchor bolts in the pump pedestal holes.

Solution:

Engineering revised the baseplate design to change the baseplate holes from rounded to slotted, to permit their alignment in the field. The design provides for shims plates to be welded in-place after the slotted holes are bolted.

Challenge 17: Increased Valve Thrust

During periodic testing, a valve operator thrust was recorded to exceed its original thrust setting.

Solution:

The original valve thrust calculations were reviewed to understand the controlling parameters and assess the effect of the observed change in thrust force. It was determined that the valve can accommodate the larger, as-found thrust. The body, the neck, the yoke, the flange, the yoke and bonnet bolts and flanges, and the operator hold-down bolts were qualified for the increased thrust. Engineering revised the valve design documents to reflect the change in thrust.

Challenge 18: Valve Safety Class Upgrade

A valve was procured as nonnuclear safety and nonseismic. It was later needed for a safety-related application.

Solution:

A commercial-grade dedication (CGD) review was performed to determine whether the valve could be used in a safety-related seismic application. It was determined that the valve body could be used, but not the operator. The operator, an active component, lacked the environmental and seismic qualification. The plant procured, installed, and preoperationally tested a new safety-related valve operator, which was environmentally and seismically qualified in accordance with the IEEE and ASME-QME standards. The plant design documents were revised to reflect the qualification and safety classification of the valve. The analysis and qualification records of the piping system and its supports were revised to reflect the new valve and operator weight and natural frequency. Importantly, the CGD was prepared by technical subject matter experts, who understood the hardware as well as the paperwork to be completed.

Challenge 19: Damaged Bolt Hole

Maintenance requested a design change to repair damaged hold-down bolt holes in the pedestal of a nonsafety and nonseismic fan.

Solution:

Engineering provided a design using commercially available SS coil-shaped threaded inserts to repair damaged bolt holes. These coil parts are inserted into the enlarged damaged bolt or stud hole, and they permit the reuse of the same bolt. Since the application is nonsafety and nonseismic, a commercial-grade material can be used, but following written plant procedures and criteria.

Challenge 20: In-Service Bolt Replacement

Maintenance requested Engineering to evaluate the possibility of replacing pump bolts during operation on a nonsafety system in the turbine building.

Solution:

Engineering analyzed the stability and strength of the pump anchorage with a single bolt removed out of the total of 16 bolts. It was concluded that one bolt could be removed without adversely affecting the operability or qualification of the pump. Therefore, with all the necessary safety precautions in place and following a step-by-step procedure for this type of change-out, Engineering authorized the bolt replacement, one at a time.

Challenge 21: Refinished Flange Face

Engineering is requested to provide guidance to restore a flange face scarred by a steam leak.

Solution:

The surface finish, striations, on a flange face are a function of the type of gasket used and are addressed in ASME B16.5. Small scratches on the surface should not be ignored, as they could easily lead to a leak. Engineering issued a design change to permit weld buildup followed by re-machining of the flange faces to the right surface finish. This is a classic repair technique, with more details available, for example, in ASME standard PCC-2 "Repair of Pressure Equipment and Piping."

Challenge 22: Parking of Hoist Assemblies

During refueling operations, the plant uses radial arm hoist assemblies (RAHAs) mounted on top of the reactor vessel head at the control rod drive mechanism (CRDM) platform beam to de-tension and re-tension the vessel head studs (Figure 7.3). Engineering was asked to develop a design to permit parking and securing the RAHAs during power operation when they are not used.

Solution:

Engineering provided the drawings and procedures to park and secure the RAHAs while the plant is in operation. The new RAHAs bracing structure was seismically designed to prevent failure that would cause adverse seismic interactions, and the bracing was designed to withstand the loss of coolant accident loads as well. A list of components to be installed and the installation steps were developed. The design allows for disconnecting the RAHAs seismic supports from the RAHAs monorail during outages and for securely parking the RAHAs by readily reconnecting the supports in preparation for power operation.

Figure 7.3 Radial arm hoist assemblies mounted on top of a reactor vessel.

Challenge 23: The Vibration Monitoring Program

Part of the plant predictive maintenance program is the vibration monitoring of rotating equipment. The periodic vibration testing program indicated that a control room filtration fan was too noisy and started to show signs of vibration.

Solution:

Engineering provided a design adding gusset plates to the fan motor base to reduce motor vibration levels. The gusset plates added stiffness to the motor support structure in the vibrating direction. The modification implemented in the field also repaired a fatigue-induced hairline crack in a weld.

Challenge 24: Seismically Qualified Fish Barrier

A population of small fish took residence in a plant intake pond that is the source to the cooling water pumping station.

Solution:

Engineering installed a fish barrier system to prevent the fish from entering the pond and reaching the cooling pumps intake structure. A floating boom and associated attachments were installed. A fishnet suited for the particular fish population was placed at the entrance to the pond. The new system of booms and nets was restrained to withstand wind and wave action, and the effects of a postulated earthquake.

Challenge 25: Major Project, SG Replacement

The challenges discussed so far have been manageable by an assigned team from plant staff. Once in a while, a plant faces a challenge so large and complex that it involves everyone on site and additional experts from industry, all collaborating in a long, sustained effort. This is what many PWRs had to face in replacing their SG.

Solution:

The replacement of SG was a multiyear effort that involved hundreds of engineering decisions, and merits a textbook of its own. However, many challenges faced can be grouped into general categories:

- Process considerations: Thermohydraulics design of the new SG by analysis and scale tests for all operating and accident modes, and thermohydraulic design details to prevent future failures, SG fluid chemistries, instrumentation and controls, etc.
- Material considerations: Failure analysis of previous SG, material selection for the replacement shells, tubes, plates, internals, base metals, weld metals, bimetallic joints, heat treatments, etc., to prevent future failures; new chemistry for primary and secondary side to prevent future tube failures; passivation treatments; electropolishing; chrome plating; copper removal; pH control agents; hydrazine additions; material suppliers; quality controls, etc.
- Code design considerations: Layout and supports design, applied loads for normal operating modes and postulated accidents, improved support arrangement such as removing SG snubbers and rerouting systems, stress analysis, ASME III Code compliance, etc.

- Fabrication and construction considerations: Cutting the connections and removing the existing SG with all the challenges this entails for a confined space, decontamination, and exiting the SG from containment, selecting qualified suppliers for the new SG, qualifications of suppliers procedures and personnel, bidding process, shop fabrication, shop examination and testing of the SG, cleaning and packaging, shipment, rigging, lifting and introduction into containment, erection in place, tie-ins, field examinations, leak and pressure tests, completion of ancillary systems, connection, and verification of instruments and controls, etc.
- Start-up testing of the systems, including vibration testing, calibration and verification of instruments and controls from the control room, etc.
- Monitoring of the SG performance during start-up and in-service, both short-term and long-term.

Challenge 26: Major Project, Reactor Pressure Vessel Head Replacement

Engineering developed a project to replace the reactor pressure vessel (RPV) head. This was a generic industry-wide initiative to prevent primary water stress corrosion cracking (PWSCC) of Ni alloy 600 base metal and welds, and subsequent wastage of the carbon steel heads by primary water boric acid corrosion.

Solution:

A new design to replace existing RPV was initiated and was implemented in an outage. The primary purpose of the change is to replace the parts and joints made of Ni alloy 600 by a material more resistant to PWSCC. The industry had spent years developing the alternate designs and materials, resulting in the selection of Alloy 690, and decided to replace the threaded head adapter connections with full penetration welds. Reactor head penetrations that were not needed were eliminated. In addition, the CRDMs were upgraded to one-piece pressure housings, which eliminated potential leakage paths that existed with the current three-piece pressure housing design.

Challenge 27: The Tornado Missile

Engineering was asked to evaluate the potential damage from a tornado missile, in this case a car, projected against an exhaust pipe.

Solution:

Part of the design of nuclear power plants is to assure that a postulated tornado will not damage essential SSCs. The tornado evaluation must address wind effects and tornado missile effects. The type, size, and velocity of tornado missiles are specified by regulation (US NRC Standard Review Plan Sections 3.3 and 3.5, and Regulatory Guide 1.76). The analysis of these effects is accomplished using a combination of analytical closed-form solutions, experimental-based parametric solutions, and numerical solutions.

The quantitative analysis of the effect of tornado missiles on SSCs dates back to the 1950s and 1960s (for example, "Fragment Effects on Structural Elements" by Holmes and Narver, and "Missile Generation and Protection in Light-Water-Cooled Power Reactor Plants" by Oak Ridge National Laboratory). Today, the analysis option can be accomplished using finite element analysis (FEA), and the qualification criteria can be based on stress or strain limits, ductility limits, or deformation limits. Engineering performed a dynamic plastic FEA analysis of such a strike from different angles (Figure 7.4) with the constriction of flow through the stack being of

Figure 7.4 Finite element analysis of tornado-born missile strike on an exhaust pipe.

primary importance. The FEA results, together with experimental data on flow constriction, were then used to assess (1) whether the pipe will bend and buckle or instead break and tear, depending on the strain level, and (2) the extent of constriction of the flow area if the pipe does not break.

7.5 STRUCTURAL

Challenge 1: Pre-staged Bolt Removal

The actuators on a main steam isolation valve (MSIV) were planned for replacement during an upcoming outage. Because of the large size of the operators, it was necessary to remove the missile shield cover over the MSIV room. Engineering was asked whether portions of the shield removal can be staged before starting the outage.

Solution:

The missile shield cover is held in position by 30 anchor bolts. A structural analysis was conducted to determine whether some bolts can be removed while still maintaining the shielding capacity of the cover. The analysis determined that missile protection can be achieved, with margin, with as little as 10 bolts so that 20 of the 30 bolts can be removed ahead of the outage. Based on this margin, Engineering authorized the pre-outage untightening of some of the bolts.

Challenge 2: Inquisitive Attitude

A 5/8-inch-diameter concrete expansion anchor bolt in a 2-feet-thick ceiling appeared not to be sufficiently spaced from two $1^1/_2$ -inch diameter through bolts. The 5/8-inch-expansion anchor bolt is used for rigging, while the $1^1/_2$ -inch bolts are for a baseplate in a room above the ceiling.

Solution:

The bolt pattern was analyzed for the maximum loads that can be imparted for the rigging operation and for the support above the ceiling. The criteria of ACI-349 were used to determine the extent of overlap of the anchor bolt cones, and, if there was overlap, to reduce the concrete anchor bolt capacity accordingly. In this case, it was determined that despite the apparent closeness of the bolts the cones do not overlap; each bolt maintains its full capacity. This is an example of a common occurrence, whereas a field observation raises a question, and the question is quantitatively evaluated to determine whether or not it is an issue that requires corrective action. Nuclear power plant staff are encouraged to have such an inquisitive mind and a questioning attitude.

Challenge 3: Hot Concrete Penetration

An accidental valve alignment resulted in 330 °F feedwater to flow through a containment penetration that was designed for normal containment ambient conditions, not for hot service. Unlike penetrations for hot service, this penetration does not have air gaps to provide cooling between the pipe and the concrete.

Solution:

The overtemperature condition was analyzed for its effect on the concrete, and the concrete was closely inspected. Because the overtemperature condition lasted for a short time, it had no time to affect the concrete. This was confirmed by the inspections.

Challenge 4: New Lift

A main steam strainer had to be lifted, disassembled for inspection, reworked if necessary, and reinstalled. A new lifting crane assembly was to be designed and shop fabricated. A nearly identical tool had been previously used.

Solution:

The designed lifting crane was analyzed for structural adequacy in accordance with the ANSI B30 series standards and the AISC Manual of Steel Construction. The ANSI B30 standards are a series of nearly 30 standards that govern the design, construction, operation, inspection, and maintenance of various types of cranes, hoist, lifting systems, and derricks. In addition, the crane was verified to meet OSHA 29 CFR 1910 requirements for lifting cranes. Once fabricated, the crane was tested for a design load exceeding the steam strainer's weight.

Challenge 5: Lifting Pads

Welded pads had to be added to existing floor beams in the turbine building. The pads were to be used to support the lift of large valves during an upcoming outage.

Solution:

Engineering retrieved the turbine building structural calculations to determine if the additional weight of the lifted valves could be accommodated, with margin. It was determined that up to 2000 lbs could be lifted, which is twice what is needed to lift one valve at a time, including the ANSI B30 safety factors on rated load. Step-by-step instructions were written for rigging the valves, one at a time.

Challenge 6: Lifting Traveling Screens

A new spreader was needed to lift the traveling screens at a river intake pump structure.

Solution:

A spreader bar was designed, analyzed, and qualified to lift up to 25,000 lbs, above what is needed to lift the screens. The existing structure was inspected, analyzed, and qualified for the new lift load. The lifting assembly was fabricated and successfully used.

Challenge 7: Reusing Abandoned Bolts

Plant staff wanted to use existing abandoned expansion anchor bolts as rigging points to lift a valve actuator during an upcoming outage.

Solution:

The design documents of the existing bolts were retrieved and confirmed that the bolts can sustain the new load. The bolts were inspected for signs of damage to the bolt and concrete. The embedment depth of the bolts was confirmed by ultrasonic inspection. The bolts were tightness checked for the installation torque. Finally, each bolt was load tested, prior to use.

Challenge 8: Anchorage for a Fall Protection System

Engineering was asked to review whether an existing anchor point is capable of sustaining a 5000-lb load for personnel fall protection, following OSHA 1926.502.

Solution:

OSHA 1926.502 requires that fall-arrest systems be capable of limiting the fall of a person to 6 ft and bring the person to a "complete stop and limit maximum deceleration distance an employee travels to 3.5 feet (1.07 m)." The personnel fall protection system was confirmed to meet the OSHA requirements, and a load well below 5000 lb was necessary, but the 5000-lb design was kept for added safety.

Challenge 9: Climbing

Engineering was asked to determine acceptable general conditions for climbing on equipment inside containment and to specify what types of equipment should not be climbed on.

Solution:

Engineering performed a review of nuclear plant practices and OSHA requirements for climbing on equipment. The review was supported by calculations and best practices. It was concluded that climbing on equipment should not be permitted, and climbing on anything other than purpose-designed ladders and platforms should be avoided.

Challenge 10: Fall Arrestor

Engineering was asked to review the safety of a new fall-arrest device that can be clamped to ladder systems, to be used with full-body harness.

Solution:

Commercially available systems, load rated and compliant to OSHA and rating standards are available. They incorporate a metal clamp that locks onto the rail and glides smoothly along the rails for hands-free operation. The assembly provides the connection between the full-body harness and the guide rail, ensuring safety during descent and ascent.

Challenge 11: Door Window

Plant staff expressed concern about a solid plant door that has the potential to hit someone on the other side on opening.

Solution:

Engineering designed and the plant installed a coated, 3-h fire-rated door, which contains a fire resistance-rated glass window. At the same time, all doors were reviewed for the same concern.

Challenge 12: Vibrating Fan

After years of reliable service, the predictive maintenance vibration monitoring system has indicated an increase in the vibration trend of an exhaust fan.

Solution:

The foundation anchor bolts were found to have loosened over the years and were revised to provide vibration dampers. While the dampers eliminated the vibration of the fan, they weakened the lateral seismic response of the fan unit and it became necessary to add lateral seismic stops (bumpers) at the base of the fan. These lateral bumpers are called snubbers by their manufacturers.

Challenge 13: New Rack Seismic Interaction

A new storage rack must be installed in a safety-related seismic category I building. The rack itself does not have a seismic safety function, but it could be an adverse seismic interaction with nearby safety-related SSCs.

Solution:

Engineering inspected the future rack location and confirmed the potential for seismic interaction. Because of the concern with seismic interaction with nearby safety-related SSCs, the rack has to be seismically restrained. The inspection provided the data necessary to seismically design the base anchorage to the floor and the lateral braces tying the rack upper-level back to the wall.

Challenge 14: Swinging Light Fixtures

Pendant light fixtures inside containment are swinging in the airflow from the discharge of an HVAC duct louvers.

Solution:

Engineering developed a design using 1/8″ SS cable to secure the light fixtures to a nearby steel structure, eliminating the swing.

Challenge 15: Gas Bottle Restraints

It is common practice in nuclear power plants to seismically restrain all equipment, whether temporary or not, inside safety-related buildings, including the containment building. Maintenance requested a clear guide of do's and don'ts when restraining compressed gas bottles.

Solution:

Engineering developed illustrated guidance on restraining compressed gas bottles. The guidance includes the following considerations:

- The bottles should be restrained at a minimum of two points, using load-rated straps or chains.
- The two restraint points should be located at least 1 foot apart with at least one above and one below the bottle midheight.
- Portable racks used for mounting compressed gas bottles that do not meet certain criteria for stability should be restrained against toppling. Once in position, carts should be prevented from rolling by using either wheel breaks or restraints.

Challenge 16: Use of Grating in Place of Concrete Plugs

Concrete floor plugs were pulled to access and inspect piping underneath the deck floor. Engineering was requested to allow grating to be installed instead of reinstalling the concrete floor plugs. This is to simplify future inspections of the piping at the same location.

Solution:

The effect of replacing the concrete floor plugs with grating platforms was investigated. The grating would consist of hinged sections for easy access to piping. The investigation addressed several aspects of the replacement, including the effect of the change on the flood analyses, on heat loads in the affected rooms, on possible corrosion mechanisms, on seismic interactions, etc. In this case, the replacement of the concrete plugs by steel grating was found to be acceptable.

Challenge 17: The Rusted Handrail

The handrail in a water intake structure was found to be rusted.

Solution:

Engineering evaluated three options: (1) clean and repaint the existing handrail, (2) replace the handrail in kind, and (3) replace the handrail with a new material. In this case, option (3) was selected, using a fiberglass-reinforced composite, compliant to OSHA 29 CFR 1910.

Challenge 18: The Rusted Stairs

Outdoor steel stairs were found to be rusted.

Solution:

The damaged sections were replaced. Maintenance added corrosion-resistant coating to the top and bottom of the stairways.

Challenge 19: Spalling of Concrete Pad

Fuel assemblies are brought to the site and into the fuel building by truck. After years of loading–unloading the fuel shipping containers, the 8-inch-thick concrete pad started to spall.

Solution:

The concrete spalling was superficial and was repaired following the plant repair procedures using an epoxy grout rather than cement grout.

Challenge 20: Postaccident Debris Accumulation

An industry-wide effort was implemented in response to the NRC's Generic Letter (GL) 2004-002. This GL addressed the potential for jet impingement from a postulated pipe break that would blow off insulation. In the worst case scenario, this blown-off insulation would form debris that would accumulate at the bottom of the containment sump and potentially reduce or block the water recirculation flow from the sump.

Solution:

The resolution of this postulated issue involved an industry-wide co-ordinated effort. Jet impingement testing was conducted to quantify the extent of blow-off of insulation from jet impingement for different types of insulation and jackets. To illustrate the depth to which the industry pursued this issue consider this: In one case, wire mesh on door signs were deemed to be potential debris and had to be removed plant-wide; these were replaced by metal clips in conjunction with cable ties made of 1//8″ SS aircraft cable.

Challenge 21: Accidental Drop during Heavy Lift

Replacing very large equipment in nuclear power plants requires lifting the equipment with a 300-ton mobile crane and transporting it to a dismantling location at the plant. In addition to the personnel safety concerns to be addressed, postulated accident scenarios are considered. For example, the following question was posed: What if the heavy equipment dropped at a ground location under which is buried safety-related pipe or duct?

Solution:

In this case of the drop of a heavy weight on the ground surface, there are good theoretical methods to conservatively predict the pressure that would be transferred from the impact at the ground surface to the buried commodity. There are also good criteria to address the next step: Whether the pressure on the buried commodity would cause damage? These studies were performed and used to decide on a case-by-case basis whether an accidental drop would be detrimental, in which case the route was avoided.

Challenge 22: Concrete Cracks

During an inspection in the turbine building, the inspectors noted cracks in the concrete pedestal of a pressure vessel.

Solution:

Cracks in concrete have been the subject of numerous engineering studies, tests, and investigations such as those conducted by the American Concrete Institute. There are reliable techniques to characterize crack shape and depth, to analyze their effect on strength, and to repair the cracks. In this particular case, the crack was measured to be shallow. They had no detrimental effect on the corrosion life of the reinforced concrete pedestal and had no effect on its strength and seismic resistance. However, as an additional precaution, the cracks were repaired by sealing and a periodic inspection was added for these locations.

Challenge 23: Looking Beyond the Original Concern

A new nonsafety-related, nonseismic auxiliary power generator had to be installed. There was an initial concern that it would have the potential to overturn during an earthquake and cause a seismic interaction with safety-related structures and equipment.

Solution:

Given the generator weight and seismic ground acceleration, Engineering determined that its aspect ratio (height/width) was sufficiently small to not constitute a seismic overturning concern. However, beyond the original seismic concern, it became evident that unless it was anchored, the generator could become a missile in case of high winds, and definitely in case of tornado. For this reason, the generator was tied down.

Challenge 24: Access Gate

The railing for the access gate to a ladder had to be replaced.

Solution:

The ladder was placed out of service, while the gate rail was removed for replacement. OSHA regulation requires ladders in the plant to have a gate access guarded by a railing. The gate and railing have explicit design characteristics.

7.6 INDUSTRY OPERATING EXPERIENCE

What is meant by IOE?

The term IOE refers to several databases of practical nuclear power plant challenges and solutions faced by engineers at operating plants. The IOE captures events that occur at nuclear power plants around the country and worldwide. The IOE consists of several formal industry databases such as INPO's Consolidated Events Database, as well as EPRI, NEI reports.

Another critical source of knowledge comes from NRC communications. These can take several forms: administrative letters, circulars, generic letters (Annex 1), bulletins (Annex 2), information notices (Annex 3), information assessment team advisories, regulatory issue summaries, and security advisories.

From this collective experience, each plant can extract lessons learned and strive for continuous improvements in safety and reliability. The nuclear power industry does not hesitate to formally share lessons learned, and workshops are conducted nationally and internationally on the subject of lessons learned in nuclear power plant engineering, maintenance, and operations.

How do the various departments use the IOE?

- Engineering uses the IOE to learn about causes and modes of failures, how to perform root cause analysis, and how to develop better designs in support of plant operation.
- Operations uses the IOE to improve plant performance and to prevent transient events and unanticipated plant perturbations.
- Maintenance uses the IOE in pre-job briefs to identify unforeseen issues that may arise during the implementation of the maintenance activity.
- Training uses the IOEs during their preparation of training classes, normally as introduction material to highlight the purpose and importance of the training.
- Licensing checks license-related IOEs to determine whether they affect the plant's TS, the final safety analysis report (FSAR), or other licensing-basis documents.

What if an event is critical, complex, or so broad that it does affect many plants?

If an event or an issue is complex or broad, then organizations such as INPO, EPRI, Owner's Groups, and NEI would get involved to study and resolve the issue generically. The objective of this sharing of knowledge is quite simple: to learn from each other in order to improve plant safety and performance. In critical cases, INPO would document and report the condition by issuing a Significant Operating Experience Report (SOER) or an Industry Event Report (IER) which describes the event, its causes and contributing factors, and recommendations. These recommendations could go beyond the scope of the original SOER and IER. Plants are expected to evaluate the report recommendations against their current programs and establish plans to close any gaps identified. Each plant would then complete and submit a formal response to INPO by an established date.

Is there another industry-wide effort intended to improve plant performance?

An important improvement program that comes to mind is what is called today the NRC Maintenance Rule (MR). The key objective of the MR is to identify preventable near misses or failures. This concept requires the identification of incorrect maintenance procedures, personnel error, or missed opportunities to prevent challenges to the plant. This is in contrast to failures due to materials, design, fabrication shortcomings, which are handled separately under the corrective action program.

Why the focus on maintenance?

In the early 1960s and 1970s, the availability (capacity factor) of some nuclear power plants (the percent of time they were online producing power) was not very effective. In addition to the losses in power production, the frequent unplanned scrams from power caused pressure and temperature transients that challenged the equipment. The industry, and the NRC, put a lot of effort in understanding the causes of such low availability. These studies pointed out that nearly one-third of these unforeseen events were attributed to maintenance shortcomings. The NRC issued an MR for plants to control the effectiveness of their maintenance program and to provide the platform for the regulator to monitor the effectiveness of plant maintenance programs. Today, nuclear power plants' capacity factor averages above 90%.

Is the MR mandatory?

The MR is mandated through 10 CFR 50.65. The MR brings in a number of documents, requirements, and guides to help the plants implement the program. These documents are available through a dedicated Web page on the NRC Web site. These documents include Regulatory Guide 1.160 "Monitoring the Effectiveness of Maintenance at Nuclear Power Plants," which describes one acceptable method for implementing the MR, and NUMARC 93-01 "Industry Guideline for Monitoring the Effectiveness of Maintenance at Nuclear Power Plants," developed with the assistance of the NEI MR Task Force, with input from EPRI, INPO, and others.

How do plants implement the MR?

Many plants implement the MR through the system engineers, who are responsible for monitoring plant systems parameters, status, preventive maintenance, and trending the systems' performance. Several tools are available for plant personnel to achieve these activities. INPO, EPRI, and NEI collaborated to develop NUMARC 93-01. There are quantitative performance measures used to evaluate essential SSCs. The performance criteria address reliability, availability as well as failure consequence. On the basis of risk (likelihood and consequence of failures or malfunctions), the plant develops corrective actions (measurements, data gathering, and monitoring) to guard against failure.

Is failure mode and effects analysis different from the MR?

Failure mode and effects analysis (FMEA) is an engineering tool used across industries, which had its infancy in the automotive industry. Its intent is to search for failure causes and failure modes of SSCs (what would cause the SSC to fail and how it would fail) and the effects (the consequences) of such failure modes. FMEA is used, for example, to evaluate whether a change to an SSC could introduce a new failure mode, one which was not part of the original design.

Where to find guidance for performing an FMEA?

Because FMEA is a technique used in most industries, there are many good textbooks and standards on the subject. For nuclear power plants, a good starting point would be EPRI NP-6406 "Guidelines for the Technical Evaluation of Replacement Items in Nuclear Power Plants (NCIG-11)" and INPO's "Equipment Performance Information Exchange." In addition, failure modes are addressed in EPRI performance databases, industry IOEs, and NRC reports.

How do plants implement FMEA?

FMEA is usually an integral part of the plant design change process and is accomplished by the engineer responsible for the design change. For more complex equipment, vendors are consulted. Plant system design criteria documents and/or FSAR would also be sources of information for FMEA.

7.7 EXAMPLES OF IOEs

IOE 1: Failure of a Lift Assembly

During a refueling outage, a temporary lift structure failed while moving a large main generator stator. The accident caused a fatality and several injuries. The impact from the fall damaged the turbine deck and electrical switchgear and caused the rupture of a fire waterline. The impact was such that a reactor coolant pump tripped on high vibration signal. The outage had to be extended, and the second unit went into a forced outage.

There were several causes for the accident, including reliance on experience with similar but not identical lifts, questions related to the structural design, and structural testing of the lift.

IOE 2: New Single-Point Vulnerability

A control circuit modification introduced a new single-point vulnerability, the possibility that a sustained loss of power to flow switches would trip the three feed pumps. After the system was placed in service with the modification, alarms indicating a loss of cooling water flow to the reactor feed pumps prompted operators to manually scram the reactor as a preemptive measure.

There were several causes for this event, including not fully addressing the loss of power scenario and the assessment that a loss of the feed pumps was an analyzed transient.

IOE 3: Yoke-to-Bonnet Stud Failure

During a refueling outage, Maintenance found a failed yoke-to-bonnet stud. Further inspection revealed a failed hand wheel key on the same valve and fatigue damage to the another stud.

The cause of this damage was traced to excessive FIV when the valve is throttled during shutdown cooling. The study of the fracture surface (fractography) confirmed the high-cycle fatigue diagnostic, with fatigue beach marks reflecting crack propagation and final fracture of the remaining metal ligament (Figure 7.5). The reader may want to refer to one of the several good textbooks on fatigue failures and fractography, including the ASM Handbook Volume 11, Failure Analysis and Prevention.

IOE 4: Feedwater Transient at Power

While operating at power, a plant feedwater pump discharge pressure dropped by approximately 40 psi at the same time as the flow decreased by approximately 450 GPM and the pump vibration velocity increased by nearly 30%. The unit was taken offline, and the feedwater pump and piping were inspected. The inspection revealed a fracture of the suction strainer upstream of the pump, which was lodged in the suction elbow to the pump.

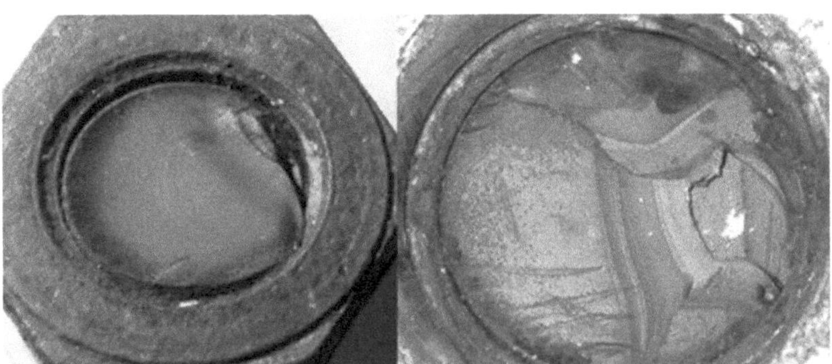

Figure 7.5 Fractography of failed valve stud.

There was no evidence of fabrication defects in the strainers; instead, the fracture surfaces showed evidence of fatigue failure. The investigation also uncovered that this was a new strainer, with different internals and wire mesh for which there was little operating experience. In addition, because the new strainers were larger than the earlier ones, they had to be relocated.

IOE 5: Cracking of Oil Cooler Pipe Threads

A motor bearing cooling water pipe was found to be leaking at its threaded joint, which was found to be cracked. The cause of the cracking was traced to a combination of fatigue and corrosion-induced pitting. The schedule 40 threaded pipe was replaced by a schedule 80 pipe. The corrosion was traced to rust buildup during warehousing of the pipe for several years.

IOE 6: Pinhole Leak

A small, dripping pinhole leak developed in a diesel generator heat exchanger water return line. The line is moderate energy (i.e., its operating temperature is below 200 °F and its operating pressure is below 275 psi), so the engineer applied ASME XI Code Case N-513 for assessment of the pinhole leak and qualified the leaking condition as acceptable as is till the outage, when the leak would be repaired. The leak was deemed to be a nuisance, but it had no safety or operations implications; a catch pan was placed under the leak.

A review of the Code Case N-513 assessment noted that the Code Case N-513 only applies to straight pipe, and the leak was at an elbow. Therefore, the assessment was declared to be invalid. This delay in disposition of the pinhole leak resulted in a delay in making the call of inoperability and entering a limited condition of operation, which then evolved into a Licensee Event Report (LER) and a Consequential Critical Component Failure. The leak was temporarily repaired using weld deposition.

IOE 7: Valve Seat Failure

The threaded upper seat in a feedwater-regulating valve unthreaded from the valve body, preventing the valve from closing or regulating flow. The power had to be reduced till the root cause was identified and the repair was completed.

The cause for the upper seat unthreading and becoming loose was FIV of the internals, caused by a shortcoming in the hydraulic sizing of the valve. The valve internals were replaced, and vibration monitoring was put in place. The maintenance procedures were revised to check the valve internals on a regular basis till it is well established that the FIV has been eliminated.

IOE 8: Leak at Socket Weld

During outage inspections, a small leak was identified in a reactor coolant pump vent line. The outage had to be extended to repair the leak. The leak was traced to a tight crack in a socket weld. The cause of the crack was a combination of slag inclusions in the weld and high-cycle fatigue caused by the pump operation. Two other similar welds were also found to have crack indications. The welds were repaired and replaced with welds with a 2:1 profile (weld leg twice as long along the pipe as they are deep along the socket).

IOE 9: Steam Leak

A small steam leak was detected, flowing from under the insulation, in the reactor coolant system. This forced a power reduction from 100% to 10% and entry into containment to pinpoint the source of leak and determine its root cause. An isolation valve was closed, which stopped the steam leak. The investigation revealed that a pipe support was missing, which caused the pipe to vibrate excessively.

IOE 10: Corrosion of Seismic Support

While removing a service water pump in the intake structure, Maintenance noticed that the seismic support to the pump was rusted. The intake structure is quite humid, and the new support steel was painted to protect it from rusting, but the paint had flaked in some places. All supports in the intake structure were inspected, cleaned, and repainted, and routine visual inspection of the intake structure was instituted.

IOE 11: Stratified Leakage Flow

While a PWR plant was at 100% power, a leak of about 0.2 GPM of reactor coolant system (RCS) primary water was detected in the containment sump. The source of the leak was a circumferential crack at an elbow butt weld in the 8 in sch. 140, 316 SS residual heat removal (RHR) pipe (Figure 7.6). The RHR isolation valve is a motor-operated wedge-type valve, which was closed during power operation. The sequence of events that lead to the crack and leak of RCS is quite interesting—a classic, documented in NRC Bulletin 88-08 Supplement 3.

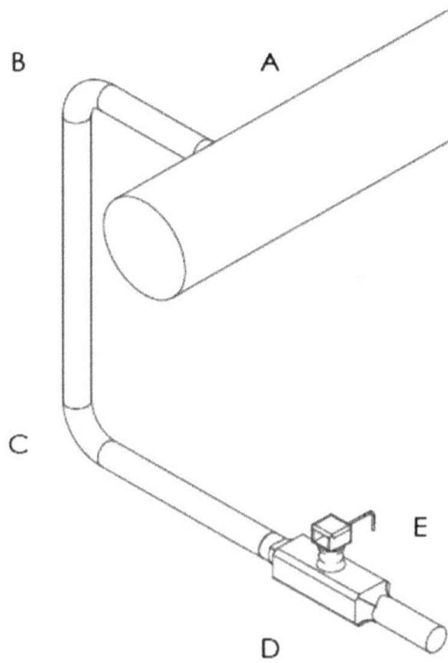

Figure 7.6 Small leak in valve packing at E causes thermal cycling in the horizontal leg C–D.

- The RCS loop operates at approximately 600 °F and 2000 psi. This pressure applies from A to D, but the temperature during normal operation drops from 600 °F at A down to the containment environment temperature, somewhere in the vertical leg B to C.
- This would be the temperature distribution if all was well. But, the valve packing gland starts to leak a very tight leak path and therefore a very low flow rate out of the packing. The leak is piped to a collection tank through the leak-off piping E.
- This small but steady leak causes the 600 °F RCS water to slowly migrate from A to B, then down the vertical leg B to C, to the valve D, and to leak-off E to a leak collection tank.
- When the 600 °F water reaches the bottom elbow C, it starts to flow horizontally, slowly, toward the leaking isolation valve, from C to D.
- Because this trickle flow is slow and much hotter than the pressurized water initially in C to D, the flow stratifies, with the 600 °F hot fluid having a lower specific gravity flowing on top of the pipe, with the stagnant cooler, heavier pressurized water staying at the bottom of the 8-inch RHR pipe.
- The temperature difference between the top and bottom causes thermal stresses in the horizontal leg as the pipe (hot on top, cooler at bottom) wants to bow upward.

- As the hot fluid reaches the valve, it heats the disk metal, which expands, sealing the valve and stopping the packing leak.
- Now, without the valve packing leak, the slow downward flow of RCS water stops, and the section B to D starts to cool down back to containment ambient temperature.
- The valve disk now cools down, contracts, reopening the leak path through the packing, and the hot stratified flow process repeats itself.
- In conclusion, we have a series of stratification heat-ups followed by cooldowns of the horizontal pipe leg C to D, which caused a complex pattern of fluctuating stresses in that leg and finally caused the weld to crack at the elbow C.

IOE 12: Thermal Sleeves

Thermal sleeves are cylinder welded inside branch connections or nozzles to reduce the thermal shock from thermal mixing of the pipe and vessel fluid streams. An example of thermal sleeve is illustrated in Figure 7.7. Thermal sleeves are installed in several systems, including the RCS branches in PWRs and boiling water reactors, because these connections must be designed to sustain thermal transients caused by the potential injection of colder water into the hot operating RCS. Some sleeves are made of SS (such as SA336 Class F8M or SA 312 TP 304); others are made of a nickel alloy (such as Inconel 600 with alloy 82 and 182 weld metal). Starting in the 1980s, several failures of thermal sleeves prompted radiographic inspections to check whether the sleeves were still in position. In one case, the thermal sleeves from all four safety injection nozzles were

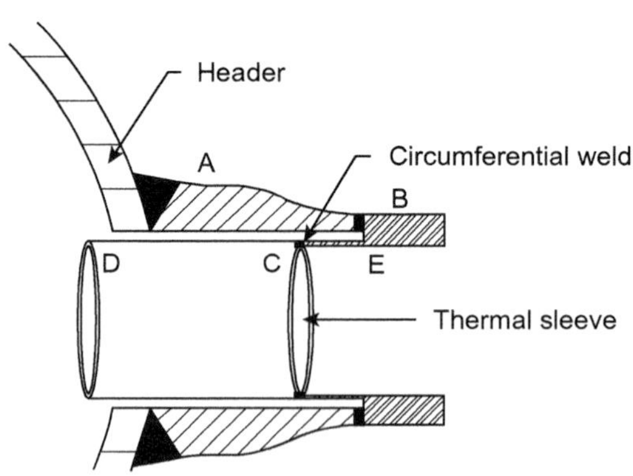

Figure 7.7 Thermal sleeve to reduce thermal stresses in the nozzle failures occurred at the weld in C or in E.

missing. In another case, 24 of 28 sleeves inspected had cracks. These failures also cause loose parts of the sleeves to be carried downstream and to get lodged inside equipment or at strainers. These failures were documented in LERs, as well as NRC Information Notice 82-30: "Loss of Thermal Sleeves in Reactor Coolant System Piping at Certain Westinghouse PWR Power Plants," Information Notice 90-10 "Primary Water Stress Corrosion Cracking (PWSCC) of Inconel 600," NUREG/CR-6260 "Application of NUREG/CR-5999 Interim Fatigue Curves to Selected Nuclear Plant Components."

The thermal sleeve failures were attributed to (1) FVI, (2) thermal cycling in the sleeve, or (3) PWSCC in the case of the Inconel 600 sleeves. The repairs consisted in (1) change of the sleeve geometry and the weld size between the sleeve and the branch or nozzle, (2) change in flow rate for those sleeves that failed from flow mixing during normal operation (usually by increasing the makeup flow rate coming through the branch line), (3) change to the material (such as switching to SA336 Class F316 or 316L welded with 308L filler metal, or Inconel 690), and (4) enhancement of the loose part monitoring program and operator awareness of this type of failure. These changes were accompanied by enhanced inspections, including borescope inspections, radiography, and ultrasonic inspection of the nozzle or branch welds near the sleeve. Volumetrically inspecting the sleeve welds is a challenge.

IOE 13: Mixing Tee

In the late 1990s, instances of high-cycle fatigue cracking at mixing tees were reported at several nuclear power plants worldwide. In one case the crack propagated through the wall and caused a leak. To understand this effect, consider the mixing that occurs in the system illustrated in Figure 7.8: Hot water flowing from A mixes at C with cooler flow coming from the outlet of a heat exchanger at B. The mixing at C causes turbulent vortices before the temperature evens and the flow continues toward D. What happens when the two liquid streams of different temperature mix at the tee in C is that the turbulent mixing of hot and cold liquid causes cyclic stresses in the pipe wall at and downstream of the mixing tee at C. Extensive numerical simulation (computational fluid dynamics coupled with heat transfer and stress FEA) and hydraulic testing have been conducted since the cracking and leaks were discovered in the late 1990s, which greatly helped understand and explain the cracking mechanism observed at the mixing tees.

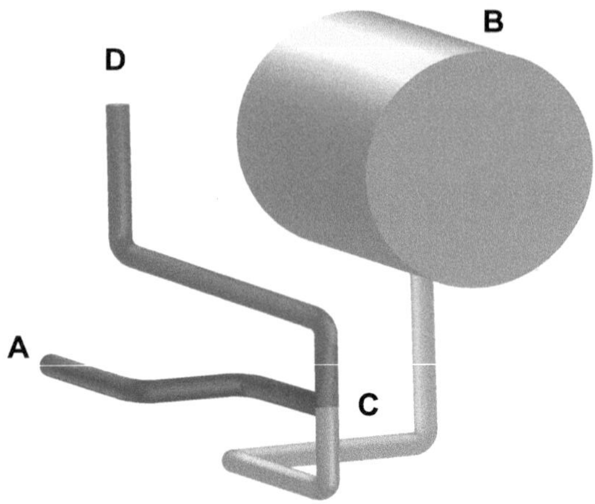

Figure 7.8 The cold stream from the heat exchanger B and the hot stream coming from A mix at the tee in C the mixed stream flows toward D cyclic thermal stresses occur in the mixing tee at C.

Analyses and testing of these mixing effects, as well as guidelines to prevent damage from these effects have been conducted by several groups and are documented in several reports, including EPRI Material Reliability Program (MRP)-192 "Materials Reliability Program Assessment of RHR Mixing Tee Thermal Fatigue in PWR Plants" and the European Commission's Joint Research Center's report EUR 22763 EN "Development of a European Procedure for Assessment of High Cycle Fatigue in Light Water Reactors: Final Report of the NESC-Thermal Fatigue." The assessment of thermal mixing can be performed at different levels of detail, depending on the severity of the mixing. In the simplest case, a threshold difference in temperature of approximately 140 °F (60 °C) between the hot and cold stream has proven to be acceptable for long-term operation. A second approach would rely on a classic linear elastic fatigue analysis, consistent with ASME III NB-3200 or NB-3600. A third approach would be based on a detailed fracture mechanics analysis of a postulated crack at the worst mixing location.

ANNEX 1

Generic Letters

Source: http://www.nrc.gov/reading-rm/doc-collections/gen-comm/gen-letters

GL-200801	Managing Gas Accumulation in Emergency Core Cooling, Decay Heat Removal, and Containment Spray Systems
GL-200701	Inaccessible or Underground Power Cable Failures that Disable Accident Mitigation Systems or Cause Plant Transients.
GL-200603	Potentially Nonconforming Hemyc and MT Fire Barrier Configurations
GL-200602	Grid Reliability and the Impact on Plant Risk and the Operability of Offsite Power
GL-200601	Steam Generator Tube Integrity and Associated Technical Specifications
GL-200402	Potential Impact of Debris Blockage on Emergency Recirculation during Design Basis Accidents at Pressurized-Water Reactors
GL-200401	Requirements for Steam Generator Tube Inspections
GL-03001	Control Room Habitability
GL-99002er	NRC Generic Letter 99-02 (Errata): Laboratory Testing of Nuclear-Grade Activated Charcoal
GL-99002	NRC Generic Letter 99-02: Laboratory Testing of Nuclear-Grade Activated Charcoal
GL-99001	NRC Generic Letter 99-01: Recent Nuclear Material Safety and Safeguards Decision on Bundling Exempt Quantities
GL-98005	NRC Generic Letter 98-05: Boiling Water Reactor Licensees Use of the BWRVIP-05 Report to Request Relief from Augmented Examination Requirements on Reactor Pressure Vessel Circumferential Shell Welds

Continued

—Cont'd

GL-98004	NRC Generic Letter 98-04: Potential for Degradation of the Emergency Core Cooling System and the Containment Spray System After a Loss-of-Coolant Accident Because of Construction and Protective Coating Deficiencies and Foreign Material in Containment
GL-98003	NRC Generic Letter 98-03: NMSS Licensees' and Certificate Holders' Year 2000 Readiness Programs
GL-98002	NRC Generic Letter No. 98-02: Loss of Reactor Coolant Inventory and Associated Potential for Loss of Emergency Mitigation Functions While in a Shutdown Condition
GL-98001	NRC Generic Letter No. 98-01: Year 2000 Readiness of Computer Systems at Nuclear Power Plants
GL-98001s1	NRC Generic Letter No. 98-01 Supplement 1: Year 2000 Readiness of Computer Systems at Nuclear Power Plants
GL-97006	NRC Generic Letter 97-06: Degradation of Steam Generator Internals
GL-97005	NRC Generic Letter 97-05: Steam Generator Tube Inspection Techniques
GL-97004	NRC Generic Letter 97-04: Assurance of Sufficient Net Positive Suction Head for Emergency Core Cooling and Containment Heat Removal Pumps
GL-97003	Annual Financial Update of Surety Requirements for Uranium Recovery Licensees
GL-97002	Revised Contents of the Monthly Operating Report
GL-97001	Degradation of Control Rod Drive Mechanism Nozzle and Other Vessel Closure Head Penetrations
GL-96007	Interim Guidance on Transportation of Steam Generators
GL-96006s1	NRC Generic Letter 96-06, Supplement 1: Assurance of Equipment Operability and Containment Integrity During Design-Basis Accident Conditions
GL-96006	Assurance of Equipment Operability and Containment Integrity During Design-Basis Accident Conditions
GL-96005	Periodic Verification of Design-Basis Capability of Safety-Related Power-Operated Valves
GL-96004	Boraflex Degradation in Spent Fuel Pool Storage Racks

—Cont'd

GL-96002	Reconsideration of Nuclear Power Plant Security Requirements Associated with an Internal Threat
GL-96003	NRC Generic Letter 96-03: Relocation of the Pressure Temperature Limit Curves and Low Temperature Overpressure Protection System Limits
GL-96001	Testing of Safety-Related Circuits
GL-950010	Relocation of Selected Technical Specifications Requirements Related to Instrumentation
GL-95009	Monitoring and Training of Shippers and Carriers of Radioactive Materials
GL-95008, Rev. 1	10 CFR 50.54(p) Process for Changes to Security Plans Without Prior NRC Approval [GL 95-08, Rev. 1 has been superseded by RIS 2014-02]
GL-95008	10 CFR 50.54(p) Process for Changes to Security Plans Without Prior NRC Approval [GL 95-08 has been superseded by RIS 2014-02]
GL-95007	Pressure Locking and Thermal Binding of Safety-Related Power-Operated Gate Valves
GL-95006	Changes in the Operator Licensing Program
GL-95005	Voltage-Based Repair Criteria for Westinghouse Steam Generator Tubes Affected by Outside Diameter Stress Corrosion Cracking
GL-95003	Circumferential Cracking of Steam Generator Tubes
GL-95004	Final Disposition of the Systematic Evaluation Program Lessons-Learned Issues
GL-95002	Use of NUMARC/EPRI Report TR-102348, "Guideline on Licensing Digital Upgrades," in Determining the Acceptability of Performing Analog-to-Digital Replacements under 10 CFR 50.59
GL-95009s1	Monitoring and Training of Shippers and Carriers of Radioactive Materials
GL-95006s1	Changes in Operator Licensing Program
GL-95001	NRC Staff Technical Position on Fire Protection for Fuel Cycle Facilities
GL-94004	Voluntary Reporting of Additional Occupational Radiation Exposure Data

Continued

—Cont'd

GL-94003	Intergranular Stress Corrosion Cracking of Core Shrouds in Boiling Water Reactors
GL-94002	Long-Term Solutions and Upgrade of Interim Operating Recommendations for Thermal-Hydraulic Instabilities in BWRs
GL-94001	Removal of Accelerated Testing and Special Reporting Requirements for Emergency Diesel Generators
GL-93008	Relocation of Technical Specification Tables of Instrument Response Time Limits
GL-93007	Modification of the Technical Specification Administrative Control Requirements for Emergency and Security Plans
GL-93006	Research Results on Generic Safety Issue 106, "Piping and the Use of Highly Combustible Gases in Vital Areas"
GL-93003	Verification of Plant Records
GL-93005	Line-Item Technical Specifications Improvements to Reduce Surveillance Requirements for Testing During Power Operation
GL-93004	Rod Control System Failure and Withdrawal of Rod Control Cluster Assemblies, 10 CFR 50.54(f)
GL-93002	NRC Public Workshop on Commercial Grade Procurement and Dedication
GL-93001	Emergency Response Data System Test Program
GL-92009	Limited Participation by NRC in the IAEA International Nuclear Event Scale
GL-92008	Thermo-Lag 330-1 Fire Barriers
GL-92007	Office of Nuclear Reactor Regulation Reorganization
GL-92006	Operator Licensing National Examination Schedule
GL-92005	NRC Workshop on the Systematic Assessment of Licensee Performance (SALP) Program
GL-92004	Resolution of the Issues Related to Reactor Vessel Water Level Instrumentation in BWRs Pursuant to 10 CFR 50.54(f)
GL-92001rs	NRC Generic Letter 92-01, Revision 1, Supplement 1: Reactor Vessel Structural Integrity
GL-92003	Compilation of the Current Licensing Basis: Request for Voluntary Participation in Pilot Program

—Cont'd

GL-92002	Resolution of Generic Issue 79, "Unanalyzed Reactor Vessel (PWR) Thermal Stress During Natural Convection Cooldown"
GL-92001r1	Reactor Vessel Structural Integrity
GL-92001	Reactor Vessel Structural Integrity
GL-91019	Information to Addressees Regarding New Telephone Numbers for NRC Offices Located in One White Flint North
GL-91018	Information to Licensees Regarding Two NRC Inspection Manual Sections on Resolution of Degraded and Nonconforming Conditions and on Operability [GL 91-18 has been superseded RIS 2005-20]
GL-91017	Generic Safety Issue 29, "Bolting Degradation or Failure in Nuclear Power Plants"
GL-91018r1	NRC Generic Letter 91-18, Revision 1: Information to Licensees Regarding NRC Inspection Manual Section on Resolution of Degraded and Nonconforming Conditions [GL 91-18 R1 has been superseded RIS 2005-20]
GL-91016	Licensed Operators' and Other Nuclear Facility Personnel Fitness for Duty
GL-91014	Emergency Telecommunications
GL-91015	Operating Experience Feedback Report, Solenoid-Operated Valve Problems at U.S. Reactors
GL-91013	Request for Info Related to the Resolution of GI 130, "Essential Service Water System Failures at Multi-Unit Sites"
GL-91010	Explosives Searches at Protected Area Portals
GL-91012	Operator Licensing National Examination Schedule
GL-91011	Resolution of Generic Issues 48, "LCOs for Class 1E Vital Instrument Buses," and 49, "Interlocks and LCOs for Class 1E Tie Breakers," Pursuant to 10 CFR 50.54
GL-91009	Modification of Surveillance Interval for the Electrical Protective Assemblies in Power Supplies for the Reactor Protection System
GL-91008	Removal of Component Lists from Technical Specifications
GL-91007	GI-23, "Reactor Coolant Pump Seal Failures" and Its Possible Effect on Station Blackout

Continued

—Cont'd

GL-91006	Resolution of Generic Issue A-30, "Adequacy of Safety-Related DC Power Supplies," Pursuant to 10 CFR 50.54(f)
GL-91005	Licensee Commercial-Grade Procurement and Dedication Programs
GL-91004	Changes in Technical Specification Surveillance Intervals to Accommodate a 24-Month Fuel Cycle
GL-91003	Reporting of Safeguards Events
GL-91002	Reporting Mishaps Involving LLW Forms Prepared for Disposal
GL-91001	Removal of the Schedule for the Withdrawal of Reactor Vessel Material Specimens from Technical Specifications
GL-90009	Alternative Requirements for Snubber Visual Inspection Intervals and Corrective Actions
GL-90008	Simulation Facility Exemptions
GL-90007	Operator Licensing National Examination Schedule
GL-90006	Resolution of Generic Issues 70, "PORV and Block Valve Reliability," and 94, "Additional LTOP Protection for PWRs"
GL-90005	Guidance for Performing Temporary Non-Code Repair of ASME Code Class 1, 2, and 3 Piping
GL-90004	Request for Information on the Status of Licensee Implementation of GSIs Resolved with Imposition of Requirements or CAs
GL-90003s1	Relaxation of Staff Position in Generic Letter 83-28, Item 2.2 Part 2, "Vendor Interface for Safety-Related Components"
GL-90003	Relaxation of Staff Position in Generic Letter 83-28, Item 2.2 Part 2 "Vendor Interface for Safety-Related Components"
GL-90002s1	Alternative Requirements for Fuel Assemblies in the Design Features Section of Technical Specifications
GL-90002	Alternative Requirements for Fuel Assemblies in the Design Features Section of Technical Specifications
GL-90001	Request for Voluntary Participation in NRC Regulatory Impact Survey

—Cont'd

GL-89001s1	NUREG-1301 - "Offsite Dose Calculation Manual Guidance: Standard Radiological Effluent Controls for Pressurized Water Reactors"
GL-89010s3	Consideration of the Results of NRC-Sponsored Tests of Motor-Operated Valves
GL-89023	NRC Staff Responses to Questions Pertaining to Implementation of 10 CFR Part 26
GL-89022	Potential for Increased Roof Loads and Plant Area Flood Runoff Depth Due to Recent Change in Probable Maximum Precipitation Criteria Developed by the National Weather Service
GL-89021	Request for Information Concerning Status of Implementation of Unresolved Safety Issue (USI) Requirements
GL-89020	Protected Area Long-Term Housekeeping
GL-89019	Request for Actions Related to Resolution of Unresolved Safety Issue A-47 "Safety Implication of Control Systems in LWR Nuclear Power Plants" Pursuant to 10 CFR 50.54(f)
GL-89017	Planned Administrative Changes to the NRC Operator Licensing Written Examination Process
GL-89018	Resolution of Unresolved Safety Issues A-17, "Systems Interactions in Nuclear Power Plants"
GL-89016	Installation of a Hardened Wetwell Vent
GL-89014	Line-Item Improvements in Technical Specifications - Removal of 3.25 Limit on Extending Surveillance Intervals
GL-89015	Emergency Response Data System
GL-89007s1	Power Reactor Safeguards Contingency Planning for Surface Vehicle Bombs
GL-89010s2	Availability of Program Descriptions
GL-89011	Resolution of Generic Issue 101 "Boiling Water Reactor Water Level Redundancy"
GL-89012	Operator Licensing Examination
GL-89013	Service Water System Problems Affecting Safety-Related Equipment
GL-89010s5	Inaccuracy of Motor-Operated Valve Diagnostic Equipment

Continued

—Cont'd

GL-89010	Safety-Related Motor-Operated Valve Testing and Surveillance
GL-89010s1	Results of the Public Workshops
GL-89009	ASME Section III Component Replacements
GL-89008	Erosion/Corrosion-Induced Pipe Wall Thinning
GL-89007	Power Reactor Safeguards Contingency Planning for Surface Vehicle Bombs
GL-89006	Task Action Item I.D.2 - Safety Parameter Display System - 10 CFR 50.54 (f)
GL-89013s1	Service Water System Problems Affecting Safety-Related Equipment
GL-89005	Pilot Testing of the Fundamentals Examination
GL-89004s1	Guidance on Developing Acceptable Inservice Testing Programs
GL-89004	Guidance on Developing Acceptable Inservice Testing Programs
GL-89003	Operator Licensing Examination Schedule
GL-89002	Actions to Improve the Detection of Counterfeit and Fraudulently Marketed Products
GL-89010s6	Information on Schedule and Grouping, and Staff Responses to Additional Public Questions
GL-89010s4	Consideration of Valve Mispositioning in Boiling Water Reactors
GL-89001	Implementation of Programmatic and Procedural Controls for Radiological Effluent Technical Specifications
GL-89010s7	Consideration of Valve Mispositioning in Pressurized-Water Reactors
GL-88020	Individual Plant Examination for Severe Accident Vulnerabilities
GL-88019	Use of Deadly Force by Licensee Guards to Prevent Theft of Special Nuclear Material
GL-88018	Plant Record Storage on Optical Disks
GL-88017	Loss of Decay Heat Removal
GL-88016	Removal of Cycle-Specific Parameter Limits from Technical Specifications

—Cont'd

GL-88015	Electric Power Systems - Inadequate Control Over Design Process
GL-88020s5	Individual Plant Examination of External Events (IPEEE) for Severe Accident Vulnerabilities - 10 CFR 50.54(f)
GL-88020s1	Initiation of the Individual Plant Examination for Severe Accident Vulnerabilities - 10 CFR 50.54
GL-88014	Instrument Air Supply System Problems Affecting Safety-Related Equipment
GL-88013	Operator Licensing Examinations
GL-88012	Removal of Fire Protection Requirements from Technical Specification
GL-88011	NRC Position on Radiation Embrittlement of Reactor Vessel Materials and Its Impact on Plant Operations
GL-88020s3	Completion of Containment Performance Improvement Program and Forwarding of Insights for Use in the IPE for Severe Accident Vulnerabilities
GL-88010	Purchase of GSA Approved Security Containers
GL-88020s4	Individual Plant Examination of External Events (IPEEE) for Severe Accident Vulnerabilities
GL-88009	Pilot Testing of Fundamentals Examination
GL-88008	Mail Sent or Delivered to the Office of Nuclear Reactor Regulation
GL-88007	Modified Enforcement Policy Relating to 10 CFR 50.49, "Environmental Qualification of Electrical Equipment Important to Safety"
GL-88020s2	Accident Management Strategies for Consideration in the Individual Plant Examination Process
GL-88006	Removal of Organization Charts from Technical Specification Administrative Control Requirements
GL-88005	Boric Acid Corrosion of Carbon Steel Reactor Pressure Boundary Components in PWR plants
GL-88004	Distribution of Gems Irradiated in Research Reactors
GL-88003	Resolution of Generic Safety Issue 93, "Steam Binding of Auxiliary Feedwater Pumps"

Continued

—Cont'd

GL-88001s1	NRC Position on Intergranular Stress Corrosion Cracking (IGSCC) in BWR Austenitic Stainless Steel Piping
GL-88001	NRC Position on IGSCC in BWR Austenitic Stainless Steel Piping
GL-88002	Integrated Safety Assessment Program II (ISAP II)
GL-87016	Transmittal of NUREG-1262, "Answers to Questions on Implementation of 10 CFR 55 on Operators' Licenses"
GL-87015	Policy Statement on Deferred Plants
GL-87014	Operator Licensing Examinations
GL-87013	Integrity of Requalification Examinations at Non-Power Reactors
GL-87012	Loss of Residual Heat Removal while the Reactor Coolant System is Partially Filled
GL-87011	Relaxation in Arbitrary Intermediate Pipe Rupture Requirements
GL-87010	Implementation of 10 CFR 73.57, Requirements for FBI Criminal History Checks
GL-87009	Sections 3.0 and 4.0 of Standard Tech Specs on Limiting Conditions for Operation and Surveillance Requirements
GL-87008	Implementation of 10 CFR 73.55 Miscellaneous Amendments and Search Requirements
GL-87007	Information Transmittal of Final Rulemaking for Revisions to Operator Licensing - 10 CFR 55 and Confirming Amendments
GL-87006	Periodic Verification of Leak Tight Integrity of Pressure Isolation Valves
GL-87005	Request for Additional Information Assessment of Licensee Measures to Mitigate and/or Identify Potential Degradation of Mark I Drywells
GL-87004	Temporary Exemption From Provisions of the FBI Criminal History Rule for Temporary Workers
GL-87003	Verification of Seismic Adequacy of Mechanical and Electrical Equipment in Operating Reactors (USI A-46)
GL-87002	Verification of Seismic Adequacy of Mechanical and Electrical Equipment in Operating Reactors (USI A-46)

—Cont'd

GL-87002s1	Transmits Supplemental Safety Evaluation Report No. 2 on SQUG Generic Implementation Procedure, Revision 2, as Corrected February 14, 1992 (GIP-2)
GL-87001	Public Availability of the NRC Operator Licensing Examination Question Bank
GL-86016	Westinghouse ECCS Evaluation Models
GL-86017	Availability of NUREG-1169, "Technical Findings Related to Generic Issue C-8, BWR MSIC Leakage and Treatment Methods"
GL-86015	Information Relating to Compliance with 10 CFR 50.49 "Environmental Qualification of Electric Equipment Important to Safety for Nuclear Power Plants"
GL-86014	Operator Licensing Examinations
GL-86013	Potential Inconsistency Between Plant Safety Analyses and Technical Specifications
GL-86012	Criteria for Unique Purpose Exemption From Conversion From the Use of HEU Fuel
GL-86011	Distribution of Products Irradiated in Research Reactors
GL-86006	Implementation of TMI Action Item II.K.3.5, "Automatic Trip of Reactor Coolant Pumps"
GL-86005	Implementation of TMI Action Item II.K.3.5, "Automatic Trip of Reactor Coolant Pumps"
GL-86010	Implementation of Fire Protection Requirements
GL-86009	Technical Resolution of Generic Issue B-59, (N-1) Loop Operation in BWRs and PWRs
GL-86010s1	Fire Endurance Test Acceptance Criteria for Fire Barrier Systems Used to Separate Redundant Safe Shutdown Trains Within the Same Fire Area
GL-86008	Availability of Supplement 4 to NUREG-0933, "A Prioritization of Generic Safety Issues"
GL-86007	Transmittal of NUREG-1190 Regarding the San Onofre Unit 1 Loss of Power and Water Hammer Event
GL-86004	Policy Statement On Engineering Expertise on Shift
GL-86003	Applications for License Amendments
GL-86002	Technical Resolution of Generic Issue B-19 - Thermal Hydraulic Stability

Continued

—Cont'd

GL-86001	Safety Concerns Associated with Pipe Breaks in the BWR Scram System
GL-85022	Potential for Loss of Post-LOCA Recirculation Capability Due to Insulation Debris Blockage
GL-85021	This Generic Communication was not issued
GL-85020	Resolution of Generic Issue 69: High Pressure Injection/ Make-up Nozzle Cracking in Babcock And Wilcox Plants
GL-85019	Reporting Requirements on Primary Coolant Iodine Spikes
GL-85018	Operator Licensing Examinations
GL-85017	Availability of Supplements 2 and 3 to NUREG-0933, "A Prioritization of Generic Safety Issues"
GL-85016	High Boron Concentrations
GL-85015	Information on Deadlines for 10CFR50.49, "Environmental Qualification of Electric Equipment Important to Safety at Nuclear Power Plants"
GL-85014	Commercial Storage at Power Reactor Sites of Low Level Radioactive Waste Not Generated by the Utility
GL-85013	Transmittal of NUREG-1154 Regarding The Davis-Besse Loss of Main And Auxiliary Feedwater Event
GL-85012	Implementation of TMI Action Item II.K.3.5, "Automatic Trip of Reactor Coolant Pumps"
GL-85011	Completion of Phase II of "Control of Heavy Loads at Nuclear Power Plants" NUREG-0612
GL-85010	Technical Specification for Generic Letter 83-28, Items 4.3 and 4.4
GL-85009	Technical Specifications for Generic Letter 83-28, Item 4.3
GL-85008	10 CFR 20.408 Termination Reports - Format
GL-85007	Implementation of Integrated Schedules for Plant Modifications
GL-85002	Staff Recommended Actions ... Regarding Steam Generator Tube Integrity
GL-85006	Quality Assurance Guidance for ATWS Equipment That is Not Safety-Related
GL-85005	Inadvertent Boron Dilution Events

—Cont'd

GL-85004	Operating Licensing Examinations
GL-85003	Clarification of Equivalent Control Capacity for Standby Liquid Control Systems
GL-85001	Fire Protection Policy Steering Committee Report
GL-84024	Certification of Compliance to 10 CFR 50.49, Environmental Qualification of Electric Equipment Important to Safety for Nuclear Power Plants
GL-84023	Reactor Vessel Water Level Instrumentation in BWRs
GL-84022	10 CFR 20.408 Termination Reports - Format
GL-84021	Long Term Low Power Operation in Pressurized Water Reactors
GL-84020	Scheduling Guidance for Licensee Submittals of Reloads That Involve Unreviewed Safety Questions
GL-84019	Availability of Supplement 1 to NUREG-0933, "A Prioritization of Generic Safety Issues"
GL-84018	Filing of Applications for Licenses and Amendments
GL-84017	Annual Meeting to Discuss Recent Developments Regarding Operator Training, Qualifications, and Examinations
GL-84015	Proposed Staff Actions to Improve and Maintain Diesel Generator Reliability
GL-84016	Adequacy of On-Shift Operating Experience for Near Term Operating License Applicants
GL-84014	Replacement and Requalification Training Program
GL-84009	Recombiner Capability Requirements of 10 CFR 50.44(c)(3)(ii)
GL-84013	Technical Specification for Snubbers
GL-84012	Compliance with 10 CFR 61 and Implementation of Radiological Effluent Tech Specs, Attendant Process Control Program
GL-84010	Administration of Operating Tests Prior to Initial Criticality
GL-84011	Inspections of BWR Stainless Steel Piping
GL-84008	Interim Procedures for NRC Management of Plant-Specific Backfitting

Continued

—Cont'd

GL-84006	Operator and Senior Operator License Examination Criteria For Passing Grade
GL-84005	Change to NUREG-1021, "Operator Licensing Examiner Standards"
GL-84007	Procedural Guidance for Pipe Replacement at BWRs
GL-84004	Safety Evaluation of Westinghouse Topical Reports Dealing with Elimination of Postulated Pipe Breaks in PWR Primary Main Loops (Generic Letter 84-04)
GL-84003	Availability of NUREG-0933, "A Prioritization of Generic Safety Issues"
GL-84002	Notice of Meeting Regarding Facility Staffing
GL-84001	NRC Use Of The Terms "Important To Safety" and "Safety Related"
GL-83040	Operator Licensing Examination
GL-83044	Availability of NUREG-1021, "Operator Licensing Examiner Standards"
GL-83043	Reporting Requirements of 10 CFR 50, Sections 50.72 and 50.73, and Standard Technical Specifications
GL-83042	Clarification to GL 81-07 Regarding Response to NUREG-0612, "Control of Heavy Loads at Nuclear Power Plants"
GL-83041	Fast Cold Starts of Diesel Generators
GL-83039	Voluntary Survey of Licensed Operators
GL-83032	NRC Staff Recommendations Regarding Operator Action for Reactor Trip and ATWS
GL-83035	Clarification of TMI Action Plan Item II.K.3.31
GL-83037	NUREG-0737 Technical Specifications
GL-83036	NUREG-0737 Technical Specifications
GL-83034	Important to Safety
GL-83038	NUREG-0965, "NRC Inventory of Dams"
GL-83033	NRC Positions on Certain Requirements of Appendix R to 10 CFR 50
GL-83028s1	Required Actions Based on Generic Implications of Salem ATWS Events

—Cont'd

GL-83031	Safety Evaluation of "Abnormal Transient Operating Guidelines"
GL-83029	This Generic Communication was not issued
GL-83023	Safety Evaluation of "Emergency Procedure Guidelines"
GL-83030	Deletion of STD. Tech Spec Surveillance Requirement 4.8.1.1.2.d.6 for Diesel Generator Testing
GL-83012a	Issuance of NRC FORM 398 - Personal Qualifications Statement - Licensee
GL-83028	Required Actions Based on Generic Implications of Salem ATWS Events
GL-83027	Surveillance Intervals in Standard Technical Specifications
GL-83026	Clarification Of Surveillance Requirements for Diesel Fuel Impurity Level Tests
GL-83025	Issuance of NRC Form 398 Personnal Qualifications Statement - Licensee
GL-83024	TMI Task Action Plan Item I.G.1, "Special Low Power Testing and Training," Recommendations for BWRS
GL-83011s1	NRC Generic Letter 83-11, Supplement 1: Licensee Qualification for Performing Safety Analyses
GL-83022	Safety Evaluation of "Emergency Response Guidelines"
GL-83021	Clarification of Access Control Procedures for Law Enforcement Visits
GL-83020	Integrated Scheduling for Implementation of Plant Modifications
GL-83019	New Procedures for Providing Public Notice Concerning Issuance of Amendments to Operating Licenses
GL-83018	NRC Staff Review of the BWR Owners' Group (BWROG) Control Room Survey Program
GL-83017	Integrity of Requalification Examinations for Renewal of Reactor Operator and Senior Reactor Operator Licenses
GL-83016a	Transmittal of NUREG-0977 Relative to the ATWS Events at Salem Generating Station, Unit NO. 1
GL-83016	Transmittal of NUREG-0977 Relative to the ATWS Events at Salem Generating Station, Unit NO. 1

Continued

—Cont'd

GL-83015	Implement. of Reg. Guide 1.150, "Ultrasonic Testing of RX Vessel Welds During Preservice & Inservice Examinations, REV.1"
GL-83014	Definition of "Key Maintenance Personnel," (Clarification of Generic Letter 82-12)
GL-83013	Clarification of Survell. Req's for HEPA Filters and Charcoal Absorber Units In STD. Tech Specs on ESF Cleanup Systems
GL-83012	Issuance of NRC Form 398 - Personal Qualifications Statement - Licensee
GL-83007	The Nuclear Waste Policy Act of 1982
GL-83011	Licensee Qualification for Performing Safety Analyses in Support of Licensing Actions
GL-83010f	Resolution of TMI Action Item II.K.3.5., "Automatic Trip of Reactor Coolant Pumps"
GL-83010e	Resolution of TMI Action Item II.K.3.5., "Automatic Trip of Reactor Coolant Pumps"
GL-83010d	Resolution of TMI Action Item II.K.3.5., "Automatic Trip of Reactor Coolant Pumps"
GL-83010c	Resolution of TMI Action Item II.K.3.5., "Automatic Trip of Reactor Coolant Pumps"
GL-83010b	Resolution of TMI Action Item II.K.3.5., "Automatic Trip of Reactor Coolant Pumps"
GL-83010a	Resolution of TMI Action Item II.K.3.5., "Automatic Trip of Reactor Coolant Pumps"
GL-83009	Review of Combustion Engineering Owners' Group Emergency Procedures Guideline Program
GL-83008	Modification of Vacuum Breakers on Mark I Containments
GL-83005	Safety Evaluation of "Emergency Procedure Guidelines, Revision 2," June 1982
GL-83004	Regional Workshops Regarding Supplement 1 to NUREG-0737, Requirements for Emergency Response Capability
GL-83006	Certificates and Revised Format for Reactor Operator and Senior Reactor Operator Licenses
GL-83003	Regulatory Guide 1.150, "Ultrasonic Testing of Reactor Vessel Welds During Pre-Service and Inservice Examinations"

—Cont'd

GL-83001	Operator Licensing Examination Site Visit
GL-83002	NUREG-0737 Technical Specifications
GL-82032	Draft Steam Generator Report (SAI)
GL-82030	Filings Related to 10 CFR 50 Production and Utilization Facilities
GL-82039	Problems with Submittals of Subsequent Information of CURT 73.21 For Licensing Reviews
GL-82038	Meeting to Discuss Developments for Operator Licensing Examinations
GL-82037	This Generic Letter was Never Issued
GL-82036	This Generic Letter was Never Issued
GL-82035	This Generic Letter was Never Issued
GL-82034	This Generic Letter was Never Issued
GL-82031	This Generic Letter was Never Issued
GL-82033	Supplement 1 to NUREG-0737 - Emergency Response Capabilities
GL-82029	This Generic Letter was Never Issued
GL-82028	Inadequate Core Cooling Instrumentation System
GL-82027	Transmittal of NUREG-0763 - Guidelines For Confirmatory In-Plant Tests of Safety-Relief Valve Discharge for BWR Plants
GL-82026	NUREG-0744, REV. 1 - Pressure Vessel Material Fracture Toughness
GL-82024	Safety Relief Valve Quencher Loads: BWR MARK II and III Containments
GL-82025	Integrated IAEA Exercise for Physical Inventory at LWRS
GL-82023	Inconsistency Between Requirements of 10CFR 73.40(D) and Standard Tech Specs For Performing Audits of Sadeguards Contin
GL-82022	Congressional Request for Information Concerning Steam Generator Tube Integrity
GL-82020	Guidance for Implementing the Standard Review Plan Rule
GL-82018	Reactor Operator and Senior Reactor Operator Requalification Examinations

Continued

—Cont'd

GL-82021	Fire Protection Audits
GL-82019	Submittal of Copies of Documentation to NRC-Copy Requirements for Emergency Plans and Physical Security Plans
GL-82017	Inconsistency of Requirements Between 50.54(T) and 50.15
GL-82016	NUREG-0737 Technical Specifications
GL-82015	This Generic Letter was Never Issued
GL-82014	Submittal of Documents to the NRC
GL-82013	Reactor Operator and Senior Reactor Operator Examinations
GL-82012	Nuclear Power Plant Staff Working Hours
GL-82011	Transmittal of NUREG-0916 Relative to the Restart of R.E. Ginna Nuclear Power Plant
GL-82010	Post-TMI Requirements
GL-82009	Environmental Qualification of Safety Related Electrical Equipment
GL-82008	Transmittal of NUREG-0909 Relative to the Ginna Tube Rupture
GL-82007	Transmittal of NUREG-0909 Relative to the Ginna Tube Rupture
GL-82006	This Generic Letter was Never Issued
GL-82003	High Burnup MAPLHGR Limits
GL-82005	Post-TMI Requirements
GL-82004	Use of INPO See-in Program
GL-82002	Commission Policy on Overtime
GL-82001	New Applications Survey
GL-81037	ODYN Code Reanalysis Requirements
GL-81040	Qualifications of Reactor Operators
GL-81039	NRC Volume Reduction Policy
GL-81038	Storage of Low Level Radioactive Wastes at Power Reactor Sites
GL-81036	Revised Schedule for Completion of TMI Action Plan Item II.D.1, Relief and Safety Valve Testing

—Cont'd

GL-81034	Safety Concerns Associated with Pipe Breaks in the BWR Scram System
GL-81035	Safety Concerns Associated with Pipe Breaks in the BWR Scram System
GL-81033	Technical Specification for Station Batteries Multiplant Action
GL-81032	NUREG-0737, Item II.K.3.44-Evaluation of Anticipated Transients Combined with Single Failure
GL-81029	Simulator Examinations
GL-81031	Small Break LOCA Confirmatory Integral Systems Experiments for B&W Designed Plants
GL-81030	Safety Concerns Associated with Pipe Breaks in the BWR Scram System
GL-81028	Steam Generator Overfill
GL-81026	Licensing Requirements for Pending Construction Permit and Manufacturing License Applications
GL-81027	Privacy and Proprietary Material in Emergency Plans
GL-81023a	INPO Evaluation Reports
GL-81025	Change in Implementing Schedule for Submission and Evaluation of Upgraded Emergency Plans
GL-81024	Multi-Plant Issue B-56 Control Rods Fail to Fully Insert
GL-81023	INPO Plant Specific Evaluation Reports
GL-81016	NUREG-0737 Item I.C.1 SER on Abnormal Transient Operating Guidelines (ATOG)
GL-81013	SER for GEXL Correlation for 8X8R Fuel Reload Applications for Appendix D Submittals of the GE Topical Report
GL-81022	Engineering Evaluation of the H.B. Robinson Reactor Coolant System Leak on 1/29/81
GL-81021	Natural Circulation Cooldown
GL-81001	Qualification of Inspection, Examination, Testing and Audit Personnel
GL-81019	Thermal Shock to Reactor Pressure Vessels

Continued

—Cont'd

GL-81020	Safety Concerns Associated with Pipe Breaks in the BWR Scram System
GL-81018	BWR Scram Discharge System-Clarification of Diverse Instrumentation Requirements
GL-81015	Environmental Qualification of Class 1E Electrical Equipment-Clarification of Staffs Handling of Proprietary Information
GL-81017	Functional Criteria for Emergency Response Facilities
GL-81011	BWR Feedwater Nozzle and Control Rod Drive Return Line Nozzle Cracking (NUREG-0619)
GL-81003	Implementation of NUREG-0313, Technical Report on Material Selection & Processing GL for BWR Coolant Press Boundary Piping
GL-81006	Periodic Updating of Final Safety Analysis Reports (FSARs)
GL-81004	Emergency Procedures And Training for Station Blackout Events
GL-81012	Fire Protection Rule (45 FR 76602, 11/19/80)
GL-81010	Post-TMI Requirements for the Emergency Operations Facility
GL-81014	Seismic Qualifications for Auxiliary Feedwater Systems
GL-81007	Control of Heavy Loads
GL-81008	Odyn Code
GL-81002	Analysis, Conclusions and Recommendations Concerning Operator Licensing
GL-81009	BWR Scram Discharge System
GL-81005	Information Regarding the Program for Environmental Qualification Of Safety-Related Electrical Equipment
GL-80113	Control of Heavy Loads
GL-80112	IEB 80-25 Operating Problems with Target Rock Safety Relief Valves
GL-80111	IEB 80-17 Supplement 4, Failure of Control Rods to Insert During a Scram at a BWR
GL-80110	Periodic Updating of FSARS
GL-80109	Guidelines for SEP Soil Structure Interaction Reviews
GL-80108	Emergency Planning

—Cont'd

GL-80107	BWR Scram Discharge System
GL-80106	Report on ECCS Cladding Models, NUREG-0630
GL-80104	Orders on Environmental Qualification of Safety Related Electrical Equipment
GL-80105	Implementation of Guidance for USI A-12, Potential for Low Fracture Toughness and Lamellar Tearing on Component Supports
GL-80103	Fire Protection - Revised Federal Register Notice
GL-80102	Commission Memorandum and Order of May 23, 1980 (Referencing IEB 79-01b Supplement 2 - q.2 & 3 - Sept 30, 1980)
GL-80101	Inservice Inspection Programs
GL-80100	Appendix R to 10 CFR 50 Regarding Fire Protection-Federal Register Notice
GL-80099	Technical Specification Revisions For Snubber Surveillance
GL-80098	IEB 80-24 Prevention of Damage Due to Water Leakage Inside Containment
GL-80097	IEB 80-23 Failures of Solenoid Valves Manufactured by Valcor Engineering Corporation
GL-80096	Fire Protection
GL-80095	Generic Activity A-10
GL-80094	Emergency Plan
GL-80093	Emergency Preparedness
GL-80092	IEB 80-21 Valve Yokes Supplied by Malcolm Foundry Company, Inc.
GL-80091	ODYN Code Calculation
GL-80090	Post-TMI Requirements, NUREG-0737
GL-80089	IEB 79-01b Supplement 3 Environmental Qualification of Class 1E Equipment
GL-80088	Seismic Qualification of Auxiliary Feedwater Systems
GL-80087	Notice of Meeting to Discuss Status of EPRI-Proposed Resolution of the USI A-12 Fracture Toughness Issue
GL-80086	Notice of Meeting to Discuss Final Resolution of USI A-12

Continued

—Cont'd

GL-80085	Implementation of Guidance From Unresolved Safety Issue A-12, "Potential for Low Fracture Toughness and Lamellar Tearing on Component Supports"
GL-80084	BWR Scram System
GL-80083	Environmental Qualification of Safety-Related Equipment
GL-80082	IEB 79-01b Supplement 2 Environmental Qualification of Class 1E Equipment
GL-80081	Preliminary Clarification of TMI Action Plan Requirements - Addendum to 9/5/80 Letter
GL-80080	Preliminary Clarification of TMI Action Plan Requirements
GL-80079	IEB 80-17 Supplement 3 Failures Revealed by Testing Subsequent to Failure of Control Rods to Insert During a Scram at a BWR
GL-80078	Mark I Containment Long-Term Program
GL-80077	Refueling Water Level
GL-80076	Notice of Forthcoming Meeting with GE to Discussed Proposed BWR Feedwater Nozzle Leakage Detection System
GL-80075	Lessons Learned Tech Specs
GL-80074	Notice of Forthcoming Meeting with Representatives of EPRI to Discuss Program for Resolution of USI A-12, Fracture Tough
GL-80073	"Functional Criteria for Emergency Response Facilities," NUREG-0696
GL-80072	Interim Criteria for Shift Staffing
GL-80071	IEB 80-20 Failures of Westinghouse Type W-2 Spring Return to Neutral Control Switches
GL-80070	IEB 80-19 Failures of Mercury-Wetted Matrix Relays in RPS of Operating Nuclear Power Plants Designed by GE
GL-80069	IEB 80-18 Maintenance of Adequate Minimum Flow Through Centrifugal Charging Pumps Following Secondary Side Helb
GL-80068	IEB 80-17 Supplement 2 Failures Revealed by Testing Subsequent to Failure of Control Rods to Insert During a Scram at a BWR

—Cont'd

GL-80067	Scram Discharge Volume
GL-80066	IEB 80-17 Supplement 1 Failure of Control Rods to Insert During a Scram at a BWR
GL-80065	Request for Estimated Construction Completion and Fuel Load Schedules
GL-80064	Scram Discharge Volume Designs
GL-80063	IEB 80-17 Failure of Control Rods to Insert During a Scram at a BWR
GL-80062	TMI-2 Lessons Learned
GL-80061	TMI-2 Lessons Learned
GL-80060	Request for Information Regarding Evacuation Times
GL-80059	Transmittal of Federal Register Notice RE Regional Meetings to Discuss Environmental Qualification of Elec. Equipment
GL-80058	IEB 80-16 Potential Misapplication of Rosemount Inc. Models 1151/1152 Pressure Transmitters with "A" Or "D" Output Codes
GL-80057	Further Commission Guidance for Power Reactor Operating Licenses NUREG-0660 and NUREG-0694
GL-80056	Commission Memorandum and Order on Equipment Qualification
GL-80055	IEB 80-15 Possible Loss of Hotline With Loss of Off-Site Power
GL-80054	IEB 80-14 Degradation of Scram Discharge Volume Capability
GL-80053	Decay Heat Removal Capability
GL-80052	Five Additional TMI-2 Related Requirements - Erata Sheets to 5/7/80 Letter
GL-80051	On-Site Storage of Low-Level Waste
GL-80050	Generic Activity A-10 BWR Cracks
GL-80049	Nuclear Safeguards Problems
GL-80048	Revision to 5/19/80 Letter on Fire Protection
GL-80047	Additional Guidance on "Potential for Low Fracture Toughness and Lamellar Tearing on PWR Steam Generator and Reactor Coolant Pump Supports"

Continued

—Cont'd

GL-80046	Generic Technical Activity A-12 Fracture Toughness
GL-80045	Fire Protection Rule
GL-80044	Reorganization of Functions and Assignments Within ONRR/SSPB
GL-80043	IEB 80-13 Cracking in Core Spray Spargers
GL-80042	IEB 80-12 Decay Heat Removal System Operability
GL-80041	Summary of Meetings Held on April 22 & 23, 1980 with Representatives of the Mark I Owners Group
GL-80040	Transmittal of NUREG-0654 "Report of the B&O Task Force" And Appropriate NUREG-0626 "Generic Eval of FW Transient and Sbl"
GL-80039	IEB 80-11 Masonry Wall Design
GL-80038	Summary of Certain Non-Power Reactor Physical Protection Requirements
GL-80037	Five Additional TMI-2 Related Requirements to Operating Reactors
GL-80036	IEB 80-10 Contamination of Non-Radioactive System and Resulting Potential for Unmonitored, Uncontrolled Release to Environment
GL-80035	Effect of a DC Power Supply Failure on ECCS Performances
GL-80034	Clarification of NRC Requirements For Emergency Response Facilities at Each Site
GL-80033	Actions Req. From OL Applicants of B&W Designed NSSS Resulting From NRC B&O Task Force Review of TMI2 Accident
GL-80032	Information Request on Category I Masonry Walls Employed by Plants Under CP and OL Review
GL-80031	IEB 80-09 Hydramotor Actuator Deficiencies
GL-80030	Clarification of the Term "Operable" As It Applies to Single Failure Criterion for Safety Systems Required by TS
GL-80029	Modifications to Boiling Water Reactor Control Rod Drive Systems
GL-80028	IEB 80-08 Examination of Containment Liner Penetration Welds
GL-80027	IEB 80-07 BWR Jet Pump Assembly Failure

—Cont'd

GL-80026	Qualifications of Reactor Operators
GL-80025	IEB 80-06 Engineering Safety Feature (ESF) Reset Controls
GL-80024	Transmittal of Information on NRC "Nuclear Data Link Specifications"
GL-80023	Change of Submittal Date for Evaluation Time Estimates
GL-80022	Transmittal of NUREG-0654 "Criteria For Preparation and Evaluation of Radiological Emergency Response Plans and Prepared"
GL-80021	IEB 80-05 Vacuum Condition Resulting in Damage to Chemical Volume Control System Holdup Tanks
GL-80020	Actions Req. From OL Applicants of NSSS Designs by W and CE Resulting From NRC B&O Task Force Review of TMI2 Accident
GL-80019	Resolution of Enhanced Fission Gas Release Concern
GL-80018	Crystal River 3 Reactor Trip From Approximately 100% Full Power
GL-80017	Modifications to Boiling Water Reactor Control Rod Drive Systems
GL-80016	IEB 79-01b Environmental Qualification of Class 1E Equipment
GL-80015	Request for Additional Management and Technical Resources Information
GL-80014	LWR Primary Coolant System Pressure Isolation Valves
GL-80013	Qualification of Safety Related Electrical Equipment
GL-80012	IEB 80-04 Analysis of a PWR Main Steam Sine Break with Continued Feedwater Addition
GL-80011	IEB 80-03 Loss of Charcoal From Standard Type II, 2 Inch, Tray Absorber Cells
GL-80010	Issuance of NUREG-0588, "Interim Staff Position on Equipment Qualifications of Safety-Related Electrical Equipment"
GL-80009	Low Level Radioactive Waste Disposal
GL-80008	IEB 80-02 Inadequate Quality Assurance for Nuclear Supplied Equipment

Continued

—Cont'd

GL-80007	Information Regarding the Program For Environmental Qualification of Safety-Related Electrical Equipment
GL-80006	Issuance of NUREG-0313, Rev 1, "Technical Report on Material Selection and Processing Guidelines for BWR Coolant Pressure"
GL-80005	IEB 79-01b Environmental Qualification of Class 1E Equipment
GL-80004	IEB 80-01 Operability of ADS Valve Pneumatic Supply
GL-80003	BWR Control Rod Failures
GL-80002	Quality Assurance Requirements Regarding Diesel Generator Fuel Oil
GL-80001	NUREG-0630 "Cladding, Swelling and Rupture - Models for LOCA Analysis"
GL-79070	Environmental Monitoring for Direct Radiation
GL-79069	Cladding Rupture, Swelling, and Coolant Blockage as a Result of a Reactor Accident
GL-79068	Audit of Small Break LOCA Guidelines
GL-79067	Estimates for Evacuation of Various Areas Around Nuclear Power Reactors
GL-79066	Additional Information Re 11/09/79 Letter on ECCS Calculations
GL-79065	Radiological Environmental Monitoring Program Requirements – Enclosing Branch Technical Position, Revision 1
GL-79064	Suspension of All Operating Licenses
GL-79063	Upgraded Emergency Plans
GL-79062	ECCS Calculations on Fuel Cladding
GL-79061	Discussion of Lessons Learned Short Term Requirements
GL-79060	Discussion of Lessons Learned Short Term Requirements
GL-79058	ECCS Calculations on Fuel Cladding
GL-79057	Acceptance Criteria for the Mark I Containment Long Term Program
GL-79056	Discussion of Lessons Learned Short Term Requirements

—Cont'd

GL-79055	Summary of Meeting Held on October 12, 1979 to Discuss Responses to IE Bulletins 79-05C and 79-06C and HPI Termination Criteria
GL-79054	Containment Purging and Venting During Normal Operation
GL-79053	ATWS
GL-79052	Radioactive Release at North Anna Unit 1 and Lessons Learned
GL-79051	Follow-up Actions Resulting from the NRC Staff Reviews Regarding the TMI-2 Accident
GL-79050	Emergency Plans Submittal Dates
GL-79049	Summary of Meetings Held on 9/18-20/79 to Discuss Potential Unreviewed Safety Question on Systems Interaction for B&W PI
GL-79048	Confirmatory Requirements Relating to Condensation Oscillation Loads for the Mark I Containment Long Term Program
GL-79047	Radiation Training
GL-79046	Containment Purging and Venting During Normal Operation – Guidelines for Valve Operability
GL-79045	Transmittal of Reports Regarding Foreign Reactor Operating Experiences
GL-79044	Referencing 6/29/79 Letter Re Multiple Equipment Failures
GL-79043	Reactor Cavity Seal Ring Generic Issue
GL-79042	Potential Unreviewed Question on Interaction Between Non-Safety Grade Systems and Safety Grade Systems
GL-79041	Compliance with 40 CFR 190, EPA Uranium Fuel Cycle Standard
GL-79040	Follow-up Actions Resulting from the NRC Staff Reviews Regarding the TMI-2 Accident
GL-79039	Transmitting Division 5 Draft Regulatory Guide and Value Impact Statement
GL-79038	BWR Off-Gas Systems – Enclosing NUREG/CR-0727

Continued

—Cont'd

GL-79037	Amendment to 10 CFR 73.55 Deferral from 8/1/79 to 11/1/79
GL-79036	Adequacy of Station Electric Distribution Systems Voltages
GL-79035	Regional Meetings to Discuss Impacts on Emergency Planning
GL-79034	New Physical Security Plans (FR 43280-285)
GL-79033	Transmitting NUREG-0576 - Security Training and Qualification Plans
GL-79032	Transmitting NUREG-0578, "TMI-2 Lessons Learned"
GL-79031	Submittal of Copies of Response to 6/29/79 NRC Request
GL-79030	Transmitting NUREG-0472, Revision 2, Draft Radiological Technical Specifications
GL-79029	Transmitting NUREG-0473, Revision 2, Draft Radiological Effluent Technical Specifications
GL-79027	Operability Testing of Relief and Safety Relief Valves
GL-79028	Evaluation of Semi-Scale Small Break Experiment
GL-79026	Upgraded Standard Technical Specification Bases Program
GL-79025	Information Required to Review Corporate Capabilities
GL-79024	Multiple Equipment Failures in Safety-Related Systems
GL-79023	NRC Staff Review of Responses to I&E Bulletin 79-08
GL-79022	Enclosing NUREG-0560, "Staff Report on the Generic Assessment of Feedwater Transients in PWRs Designed by B&W"
GL-79021	Enclosing NUREG/CR 0660, "Enhancement of on Site Emergency Diesel Generator Reliability"
GL-79020	Cracking in Feedwater Lines
GL-79019	NRC Staff Review of Responses to I&E Bulletins 79-06 and 79-06a
GL-79018	Westinghouse Two-Loop NSSS
GL-79017	Reliability of Onsite Diesel Generators at Light Water Reactors
GL-79016	Meeting Re Implementation Of Physical Security Requirements

—Cont'd

GL-79015	Steam Generators-Enclosing Summary of Operating Experience with Recirculating Steam Generators, NUREG-0523
GL-79014	Pipe Crack Study Group – Enclosing NUREG-0531 and Notice
GL-79013	Schedule for Implementation and Resolution of Mark I Containment Long Term Program
GL-79012	ATWS – Enclosing Letter to GE, with NUREG-0460, Vol. 3
GL-79011	Transmitting "Summary of Operating Experience with Recalculating Steam Generators, January 1979" NUREG-0523
GL-79010	Transmitting Regulatory Guide 2.6 for Comment
GL-79009	Staff Evaluation of Interim Multiple-Consecutive Safety-Relief Valve Actuations
GL-79008	Amendment to 10 CFR 73.55
GL-79007	Seismic (SSE) and LOCA Responses (NUREG-0484)
GL-79006	Contents of the Offsite Dose Calculation Manual
GL-79005	Information Relating to Categorization of Recent Regulatory Guides by the Regulatory Requirements Review Committee
GL-79004	Referencing 4/14/78 Letter – Modifications to NRC Guidance "Review and Acceptance of Spent Fuel Pool Storage and Handling"
GL-79003	Offsite Dose Calculation Manual
GL-79002	Transmitting Rev. to Entry Control Systems Handbook (SAND 77-1033), Intrusion Detection Handbook (SAND 76-0554), and Barrier Penetration Database
GL-79001	Interservice Procedures for Instructional Systems Development – TRADOC
GL-78042	Training and Qualification Program Workshops
GL-78041	Mark II Generic Acceptance Criteria for Lead Plants
GL-78040	Training & Qualification Program Workshops
GL-78039	Forwarding of 2 Tables of Appendix I, Draft Radiological Effluent Technical Specifications, BWR, and NUREG-0133

Continued

—Cont'd

GL-78038	Forwarding of 2 Tables Of Appendix I, Draft Radiological Effluent Technical Specifications, PWR, and NUREG-0133
GL-78037	Revised Meeting Schedule & Locations for Upgraded Guard Qualifications
GL-78036	Cessation of Plutonium Shipments by Air Except in NRC Approved Containers
GL-78035	Regional Meetings to Discuss Upgraded Guard Qualifications
GL-78034	Reactor Vessel Atypical Weld Material
GL-78033	Meeting Schedule And Locations For Upgraded Guard Qualification
GL-78032	Reactor Protection System Power Supplies
GL-78031	Notice of Steam Generator Conference Agenda
GL-78030	Forwarding of NUREG-0219
GL-78029	Notice of PWR steam generator conference
GL-78028	Forwarding pages omitted from 07/11/78 letter
GL-78027	Forwarding of NUREG-0181
GL-78021	Transmitting NUREG/CR-0181 Concerning Barrier & Penetration Data Needed for Physical Security System Assessment
GL-78026	Excessive Control Rod Guide Tube Wear
GL-78025	This Generic Letter Was Never Issued
GL-78024	Model Appendix I Technical Specifications and Submittal Schedule for BWRs
GL-78022	Revision to Intrusion Detection Systems and Entry Control Handbooks and Nuclear Safeguards Technology Handbook
GL-78020	Enclosing – "A Systematic Approach to the Conceptual Design of Physical Protection Systems for Nuclear Facilities"
GL-78019	Enclosing Sandia Report SAND 77-0777, "Barrier Technology Handbook"
GL-78018	Corrected Letter On Heavy Loads Over Spent Fuel
GL-78017	Corrected Letter On Heavy Loads Over Spent Fuel
GL-78023	Manpower Requirements For Operating Reactors
GL-78015	Request for Information on Control of Heavy Loads Near Spent Fuel

—Cont'd

GL-78016	Request for Information on Control of Heavy Loads Near Spent Fuel Pools
GL-78014	Transmittal Of Draft NUREG-0219 For Comment
GL-78013	Forwarding of NUREG-0219
GL-78012	Notice of Meeting Regarding "Implementation of 10 CFR 73.55 Requirements And Status of Research…"
GL-78011	Guidance on Spent Fuel Pool Modifications
GL-78010	Guidance On Radiological Environmental Monitoring
GL-78009	Multiple-Subsequent Actuations of Safety/Relief Valves Following an Isolation Event
GL-78008	Enclosing NUREG-0408 RE Mark I Containments, and Granting Exemption from GDC 50 and Enclosing Sample Notice
GL-78007	This Generic Letter Was Never Issued
GL-78006	This Generic Letter Was Never Issued
GL-78005	Internal Distribution Of Correspondence - Asking For Comments On Mass Mailing System
GL-78004	GAO Blanket Clearance For Letter Dated 12/09/77
GL-78003	Request For Information On Cavity Annulus Seal Ring
GL-78002	Asymmetric Loads Background & Revised Request For Additional Information
GL-78001	Correction To Letter Of 12/15/77
GL-77008	Revised Intrusion Detection Systems Handbook and Entry Control Systems Handbook
GL-77007	Reliability Of Standby Diesel Generator Units
GL-77006	Enclosing Questionnaire Relating to Steam Generators
GL-77005	Nonconformity of Addresses of Items Directed to the Office of Nuclear Reactor Regulation
GL-77004	Shipments of Contaminated Components From NRC Licensed Nuclear Power Facilities to Vendors & Service Companies
GL-77003	Transmittal of NUREG-0321, "A Study of the Nuclear Regulatory Commission Quality Assurance Program"
GL-77002	Fire Protection Functional Responsibilities
GL-77001	Intrusion Detection Systems Handbook

ANNEX 2

Bulletins

BL-12-01	Design Vulnerability In Electric Power System
BL-11-01	Mitigating Strategies
BL-07-01	Security Officer Attentiveness
BL-05-02	Emergency Preparedness and Response Actions for Security-Based Events
BL-05-01	Material Control and Accounting at Reactors and Wet Spent Fuel Storage Facilities
BL-04-01	Inspection of Alloy 82/182/600 Materials Used in the Fabrication of Pressurizer Penetrations and Steam Space Piping Connections at Pressurized-Water Reactors
BL-03-04	Rebaselining of Data in the Nuclear Materials Management and Safeguards System
BL-03-03	Potentially Defective 1-inch Valves for Uranium Hexafluoride Cylinders
BL-03-02	Leakage from Reactor Pressure Vessel Lower Head Penetrations and Reactor Coolant Pressure Boundary Integrity
BL-03-01	Potential Impact of Debris Blockage on Emergency Sump Recirculation at Pressurized-Water Reactors
BL-02-02	Reactor Pressure Vessel Head and Vessel Head Penetration Nozzle Inspection Programs
BL-02-01	Reactor Pressure Vessel Head Degradation and Reactor Coolant Pressure Boundary Integrity
BL-01-01	Circumferential Cracking of Reactor Pressure Vessel Head Penetration Nozzles
BL-97-02	Puncture Testing of Shipping Packages under 10 CFR Part 71
BL-97-01	Potential for Erroneous Calibration, Dose Rate, or Radiation Exposure Measurements with Certain Victoreen Model 530 and 530SI Electrometer/Dosemeters

Continued

—Cont'd

BL-96-04	Chemical, Galvanic, or Other Reactions in Spent Fuel Storage and Transportation Casks
BL-96-03	Potential Plugging of Emergency Core Cooling Suction Strainers by Debris in Boiling-Water Reactors
BL-96-02	Movement of Heavy Loads over Spent Fuel, over Fuel in the Reactor Core, or over Safety-Related Equipment
BL-96-01	Control Rod Insertion Problems
BL-95-02	Unexpected Clogging of a Residual Heat Removal
BL-95-01	Quality Assurance Program for Transportation of Radioactive Material
BL-94-02	Corrosion Problems in Certain Stainless Steel Packagings Used to Transport Uranium Hexafluoride
BL-94-01	Potential Fuel Pool Draindown Caused by Inadequate Maintenance Practices at Dresden Unit 1
BL-93-02s1	Debris Plugging of Emergency Core Cooling Suction Strainers
BL-93-03	Resolution of Issues Related to Reactor Vessel Water Level Instrumentation in BWRs
BL-93-02	Debris Plugging of Emergency Core Cooling Suction Strainers
BL-93-01	Release of Patients After Brachytherapy Treatment With Remote Afterloading Devices
BL-92-03	Release of Patients after Brachytherapy
BL-92-01s1	Failure of Thermo-Lag 330 Fire Barrier System to Perform its Specifiied Fire Endurance Function
BL-92-02	Safety Concerns Related to "End of Life" of Aging Theratronics Teletherapy Units
BL-92-01	Failure of Thermo-Lag 330 Fire Barrier System … Cabling in Wide Cable Trays and Small Conduits Free from Fire Damage
BL-91-01s1	Reporting Loss of Criticality Safety Controls
BL-91-01	Reporting Loss of Criticality Safety Controls
BL-90-01s1	Loss of Fill-Oil in Transmitters Manufactured by Rosemount
BL-90-02	Loss of Thermal Margin Caused by Channel Box Bow

—Cont'd

BL-90-01	Loss of Fill-Oil in Transmitters Manufactured by Rosemount
BL-89-01s2	Failure of Westinghouse Steam Generator Tube Mechanical Plugs
BL-89-01s1	Failure of Westinghouse Steam Generator Tube Mechanical Plugs
BL-89-03	Potential Loss of Required Shutdown Margin During Refueling Operations
BL-89-02	Stress Corrosion Cracking of High-Hardness Type 410 Bolting Anchor Darling S350W Swing Check Valves Similar
BL-89-01	Failure of Westinghouse Steam Generator Tube Mechanical Plugs
BL-88-10s1	Nonconforming Molded-Case Circuit Breakers
BL-88-07s1	Power Oscillations in Boiling Water Reactors (BWRs)
BL-88-11	Pressurizer Surge Line Thermal Stratification
BL-88-10	Nonconforming Molded-Case Circuit Breakers
BL-88-09	Thimble Tube Thinning in Westinghouse Reactors
BL-88-08s2	Thermal Stresses in Piping Connected to Reactor Coolant Systems
BL-88-05s2	Nonconforming Materials Supplied by Piping Supplies, Inc. … and West Jersey Manufacturing Company …
BL-88-08s1	Thermal Stresses in Piping Connected to Reactor Coolant Systems
BL-88-08	Thermal Stresses in Piping Connected to Reactor Cooling Systems
BL-88-07	Power Oscillations in Boiling Water Reactors (BWRs)
BL-88-06	Actions To Be Taken for the Transportation of Model No. SPEC 2-T Radiographic Exposure Device
BL-88-05s1	Nonconforming Materials Supplied by Piping Supplies, Inc. … and West Jersey Manufacturing Company …
BL-88-05	Nonconforming Materials Supplied by Piping Supplies, Inc. … and West Jersey Manufacturing Company …
BL-88-04	Potential Safety-Related Pump Loss

Continued

—Cont'd

BL-88-08s3	Thermal Stresses in Piping Connected to Reactor Coolant Systems
BL-88-03	Inadequate Latch Engagement in HFA Type Latching Relays Manufactured by General Electric (GE) Company
BL-88-02	Rapidly Propagating Fatigue Cracks in Steam Generator Tubes
BL-88-01	Defects in Westinghouse Circuit Breakers
BL-87-02s2	Fastener Testing to Determine Conformance with Applicable Material Specifications
BL-87-02s1	Fastener Testing to Determine Conformance with Applicable Material Specifications
BL-87-02	Fastener Testing to Determine Conformance with Applicable Material Specifications
BL-87-01	Thinning of Pipe Walls in Nuclear Power Plants
BL-86-04	Defective Teletherapy Timer That May Not Terminate Treatment Dose
BL-86-03	Potential Failure of Multiple ECCS Pumps Due to Single Failure of Air-Operated Valve in Minimum Flow Recirculation Line
BL-86-02	Static "O" Ring Differential Pressure Switches
BL-86-01	Minimum Flow Logic Problems That Could Disable RHR Pumps
BL-85-03s1	Motor-Operated Valve Common Mode Failures During Plant Transients due to Improper Switch Settings
BL-85-03	Motor-Operated Valve Common Mode Failures During Plant Transients Due to Improper Switch Settings
BL-85-02	Undervoltage Trip Attachments of Westinghouse DB-50 Type Reactor Trip Breakers
BL-85-01	Steam Binding of Auxiliary Feedwater Pumps
BL-84-03	Refueling Cavity Water Seal
BL-84-02	Failure of GE Type HFA Relays In Use In Class 1E Safety Systems
BL-84-01	Cracks in BWR Mark I Containment Vent Headers
BL-83-08	Elect. Circuit Breakers With Undervoltage Trip...in Safety-Related Applications Other Than the Reactor Trip System

—Cont'd

BL-83-07s2	Apparently Fraudulent Products Sold by Ray Miller, Inc.
BL-83-07s1	Apparently Fraudulent Products Sold by Ray Miller, Inc.
BL-83-07	Apparently Fraudulent Products Sold by Ray Miller, Inc.
BL-83-06	Nonconforming Materials Supplied by Tube-Line Facilities at Long Island City, NY, Houston,TX, & Carol Stream,IL
BL-83-05	ASME Nuclear Code Pumps and Spare Parts Manufactured by the Hayward Tyler Pump Company
BL-83-04	Failure of the Undervoltage Trip Function of Reactor Trip Breakers
BL-83-03	Check Valve Failures in Raw Water Cooling Systems of Diesel Generators
BL-83-02	Stress Corrosion Cracking in Large-Diameter Stainless Steel Recirculation System Piping at BWR Plants
BL-83-01	Failure of Trip Breakers (Westinghouse DB-50) to Open on Automatic Trip Signal
BL-82-04	Deficiencies in Primary Containment Electrical Penetration Assemblies
BL-82-03r1	Stress Corrosion Cracking in Thick-Wall, Large Diameter, Stainless Steel, Recirculation System Piping at BWR Plants
BL-82-03	Stress Corrosion Cracking in Thick-Wall, Large Diameter, Stainless Steel, Recirculation System Piping at BWR Plants
BL-82-01r1s1	Alteration of Radiographs of Welds in Piping Subassemblies
BL-82-02	IE Bulletin No. 82-02: Degradation of Threaded Fasteners in the Reactor Coolant Pressure Boundary of PWR Plants
BL-82-01r1	Alteration of Radiographs of Welds in Piping Subassemblies
BL-82-01	Alteration of Radiographs of Welds in Piping Subassemblies
BL-81-02s1	Failure of Gate Type Valves to Close Against Differential Pressure
BL-81-03	Flow Blockage of Cooling Water to Safety System Components by Corbicula Sp. (Asiatic Clam) and Mytilus Sp. (Mussel)
BL-81-02	Failure of Gate Type Valves to Close Against Differential Pressure
BL-81-01r1	Surveillance of Mechancial Snubbers

Continued

—Cont'd

BL-81-01	Surveillance of Mechancial Snubbers
BL-80-17s5	Failure of Control Rods to Insert During a Scram at a BWR
BL-80-25	Operating Problems with Target Rock Safety-Relief Valves at BWRs
BL-80-17s4	Failure of Control Rods to Insert During a Scram at a BWR
BL-80-24	Prevention of Damage Due to Water Leakage Inside Containment (October 17, 1980 Indian Point 2 Event)
BL-80-23	Failures of Solenoid Valves Manufactured by Valcor Engineering Corporation
BL-80-21	Valve Yokes Supplied by Malcolm Foundry Company, Inc.
BL-80-22	Automation Industries, Model 200-520-008 Sealed-Source Connectors
BL-80-17s3	Failures Revealed by Testing Subsequent to Failure of Control Rods to Insert During a Scram at a BWR
BL-80-19r1	Failures of Mercury-Wetted Matrix Relays in Reactor Protective Systems of...Plants Designed by Combustion Engineering
BL-80-20	Failures of Westinghouse Type W-2 Spring Return to Neutral Control Switches
BL-80-19	Failures of Mercury-Wetted Matrix Relays in Reactor Protective Systems of...Plants Designed by Combustion Engineering
BL-80-18	Maintenance of Adequate Minimum Flow Thru Centrifugal Charging Pumps Following Secondary Side High Energy Line Rupture
BL-80-17s2	Failures Revealed by Testing Subsequent to Failure of Control Rods to Insert During a Scram at a BWR
BL-80-17s1	Failure of 76 of 185 Control Rods to Fully Insert During a Scram at a BWR
BL-80-17	Failure of 76 of 185 Control Rods to Fully Insert During a Scram at a BWR
BL-80-16	Potential Misapplication of Rosemount...Models 1151 and 1152 Pressure Transmitters with Either "A" or "D" Output Codes
BL-80-15	Possible Loss of Emergency Notification System (ENS) with Loss of Offsite Power

—Cont'd

BL-80-14	Degradation of Scram Discharge Volume Capability
BL-80-13	Cracking in Core Spray Spargers
BL-80-12	Decay Heat Removal System Operability
BL-80-11	Masonry Wall Design –
BL-80-10	Contamination of Nonradioactive System and Resulting Potential for Unmonitored, Uncontrolled Release to Environment
BL-80-07s1	BWR Jet Pump Assembly Failure –
BL-80-09	Hydramotor Actuator Deficiencies
BL-80-08	Examination of Containment Liner Penetration Welds
BL-80-07	BWR Jet Pump Assembly Failure –
BL-80-06	Engineered Safety Feature (ESF) Reset Controls
BL-80-05	Vacuum Condition Resulting in Damage to Chemical Volume Control System (CVCS) Holdup Tanks
BL-80-04	Analysis of a PWR Main Steam Line Break with Continued Feedwater Addition
BL-80-03	Loss of Charcoal from Standard Type II, 2 Inch, Tray Adsorber Cells
BL-80-02	Inadequate Quality Assurance for Nuclear Supplied Equipment
BL-80-01	Operability of ADS Valve Pneumatic Supply
BL-79-26r1	Boron Loss from BWR Control Blades
BL-79-01bs3	Environmental Qualification of Class IE Equipment
BL-79-01bs2	Environmental Qualification of Class IE Equipment
BL-79-03a	Longitudinal Weld Defects in ASME SA-312 Type 304 Stainless Steel Pipe
BL-79-01bs1	Environmental Qualification of Class IE Equipment
BL-79-01b	Environmental Qualification of Class IE Equipment
BL-79-28	Possible Malfunction of NAMCO Model EA180 Limit Switches at Elevated Temperatures
BL-79-27	Loss of Non-Class-1-E Instrumentation and Control Power System Bus During Operation
BL-79-26	Boron Loss from BWR Control Blades

Continued

—Cont'd

BL-79-02r2	Pipe Support Base Plate Designs Using Concrete Expansion Anchor Bolts
BL-79-25	Failures of Westinghouse BFD Relays in Safety-Related Systems
BL-79-17r1	Pipe Cracks in Stagnant Borated Water Systems at PWR Plants
BL-79-13r2	Cracking in Feedwater System Piping
BL-79-24	Frozen Lines
BL-79-23	Potential Failure of Emergency Diesel Generator Field Exciter Transformer
BL-79-14r1s2	Seismic Analyses for As-Built Safety-Related Piping Systems
BL-79-22	Possible Leakage of Tubes of Tritium Gas Used in Time Pieces for Luminosity
BL-79-13r1	Cracking in Feedwater System Piping
BL-79-02r1s1	Pipe Support Base Plate Designs Using Concrete Expansion Anchor Bolts
BL-79-14r1s1	Seismic Analyses for As-Built Safety-Related Piping Systems
BL-79-21	Temperature Effects on Level Measurements
BL-79-20	Packaging, Transport and Burial of Low-Level Radioactive Waste
BL-79-19	Packaging of Low-Level Radioactive Waste for Transport and Burial
BL-79-18	Audibility Problems Encountered on Evacuation of Personnel from High-Noise Areas
BL-79-17	Pipe Cracks in Stagnant Borated Water Systems at PWR Plants
BL-79-16	Vital Area Access Controls
BL-79-05c	Nuclear Incident at Three Mile Island – Supplement
BL-79-06c	Review of Operational Errors and System Misalignments Identified During the Three Mile Island Incident
BL-79-15s1	Deep Draft Pump Deficiencies
BL-79-15	Deep Draft Pump Deficiencies
BL-79-14r1	Seismic Analyses for As-Built Safety-Related Piping Systems
BL-79-14	Seismic Analyses for As-Built Safety-Related Piping Systems

—Cont'd

BL-79-13	Cracking in Feedwater System Piping
BL-79-02r1	Pipe Support Base Plate Designs Using Concrete Expansion Anchor Bolts
BL-79-01a	Environmental Qualification of Class IE Equipment (Deficiencies, ASCO Solenoid Valves)
BL-79-12	Short Period Scrams at BWR Facilities
BL-79-11	Faulty Overcurrent Trip Device in Circuit Breakers for Engineered Safety Systems
BL-79-10	Requalification Training Program Statistics
BL-79-05b	Nuclear Incident at Three Mile Island – Supplement
BL-79-06ar1	Review of Operational Errors and System Misalignments Identified During the Three Mile Island Incident
BL-79-09	Failures of GE Type AK-2 Circuit Breaker in Safety Related Systems
BL-79-08	Events Relevant to Boiling Water Reactors Identified During Three Mile Island Incident
BL-79-07	Seismic Stress Analysis of Safety-Related Piping
BL-79-06b	Review of Operational Errors and System Misalignments Identified During the Three Mile Island Incident
BL-79-06a	Review of Operational Errors and System Misalignments Identified During the Three Mile Island Incident
BL-79-06	Review of Operational Errors and System Misalignments Identified During the Three Mile Island Incident
BL-79-05a	Nuclear Incident at Three Mile Island
BL-79-05	Nuclear Incident at Three Mile Island
BL-79-04	Incorrect Weights for Swing Check Valves Manufactured by Velan Engineering Corporation
BL-79-03	Longitudinal Weld Defects in ASME SA-312 Type 304 Stainless Steel Pipe Spools Mfg. by Youngstown Welding & Engineering Company
BL-79-02	Pipe Support Base Plate Designs Using Concrete Expansion Anchor Bolts
BL-79-01r1	Environmental Qualification of Class 1E Equipment
BL-79-01	Environmental Qualification of Class IE Equipment
BL-78-12b	Atypical Weld Material in Reactor Pressure Vessel Welds

Continued

—Cont'd

BL-78-14	Deterioration of Buna-N Components in ASCO Solenoids
BL-78-13	Failures in Source Heads of Kay-Ray, Inc., Gauges Models 7050, 7050B, 7051, 7051B, 7060, 7060B, 7061 and 7061B
BL-78-12a	Atypical Weld Material in Reactor Pressure Vessel Welds
BL-78-12	Atypical Weld Material in Reactor Pressure Vessel Welds
BL-78-11	Examination of Mark I Containment Torus Welds
BL-78-10	Bergen-Paterson Hydraulic Shock Suppressor Accumulator Spring Coils
BL-78-09	BWR Drywell Leakage Paths Associated with Inadequate Drywell Closures
BL-78-08	Radiation Levels from Fuel Element Transfer Tubes
BL-78-07	Protection Afforded by Air-Line Respirators and Supplied-Air Hoods
BL-78-06	Defective Cutler-Hammer Type M Relays with DC Coils
BL-78-05	Malfunctioning of Circuit Breaker Auxiliary Contact Mechanism – General Electric Model CR105X
BL-78-04	Environmental Qualification of Certain Stem Mounted Limit Switches Inside Reactor Containment
BL-78-03	Potential Explosive Gas Mixture Accumulations Associated with BWR Offgas System Operations
BL-78-02	Terminal Block Qualification
BL-78-01	Flammable Contact – Arm Retainers in G.E. CR120A Relays
BL-77-08	Assurance of Safety and Safeguards During an Emergency – Locking Systems
BL-77-07	Containment Electrical Penetration Assemblies at Nuclear Power Plants under Construction
BL-77-06	Potential Problems with Containment Electrical Penetration Assemblies
BL-77-05a	Electrical Connector Assemblies
BL-77-05	Electrical Connector Assemblies
BL-77-04	Calculational Error Affecting ...Performance of a System for Controlling pH of Containment Sump Water Following a LOCA

—Cont'd

BL-77-03	On-Line Testing of the Westinghouse W Solid State Protection System (SSPS)
BL-77-02	Potential Failure Mechanism in Certain Westinghouse (W) AR Relays with Latch Attachments
BL-77-01	Pneumatic Time Delay Relay Setpoint Drift
BL-76-08	Teletherapy Units
BL-76-07	Crane Hoist Control Circuit Modifications
BL-76-06	Diaphragm Failures in Air Operated Auxiliary Actuators for Safety/Relief Valves
BL-76-05	Relay Failures – Westinghouse BFD Relays
BL-76-04	Cracks in Cold Worked Piping at BWRs
BL-76-03	Relay Malfunctions – GE Type STD Relays
BL-76-02	Relay Coil Failures – GE Type HFA, HGA, HKA, HMA Relays
BL-76-01	BWR Isolation Condenser Tube Failure
BL-75-08	PWR Pressure Instrumentation
BL-75-07	No Title - Involves Exothermic Reaction in Radwaste Shipment
BL-75-04b	Cable Fire at Browns Ferry Nuclear Power Station
BL-75-06	Defective Westinghouse Type OT-2 Control Switches
BL-75-05	Operability of Category I Hydraulic Shock and Sway Suppressors
BL-75-04a	Cable Fire at Browns Ferry Nuclear Power Station
BL-75-04	Cable Fire at Browns Ferry Nuclear Power Station
BL-75-03	Incorrect Lower Disc Spring and Clearance Dimension in Series 8300 and 8302 ASCO Solenoid Valves
BL-75-02	Defective Radionics Radiographic Exposure Devices and Source Changers
BL-75-01a	Through-Wall Cracks in Core Spray Piping at Dresden-2
BL-75-01	Through-Wall Cracks in Core Spray Piping at Dresden-2
BL-74-10b	Failures in 4-Inch Bypass Piping at Dresden 2

Continued

—Cont'd

BL-74-10a	Failures in 4-Inch Bypass Piping at Dresden 2
BL-74-16	Improper Machining of Pistons in Colt Industries (Fairbanks-Morse) Diesel-Generators
BL-74-15	Misapplication of Cutler-Hammer Three Position Maintained Switch Model No. 10250T
BL-74-14	BWR Relief Valve Discharge to Suppression Pool
BL-74-13	Improper Factory Wiring on General Electric Motor Control Centers at Fort Calhoun
BL-74-12	Incorrect Coils in Westinghouse Type SG Relays at Trojan
BL-74-11	Improper Wiring of Safety Injection Logic at Zion 1 & 2
BL-74-10	Failures in 4-Inch Bypass Piping at Dresden 2
BL-74-09	Deficiency in General Electric Model 4KV Magne-Blast Breakers
BL-74-08	Deficiency in the ITE Molded Case Circuit Breakers, Type HE-3
BL-74-07	Personnel Overexposure - Irradiation Facility
BL-74-06	Defective Westinghouse Type W-2 Control Switch Component
BL-74-04a	Malfunction of Target Rock Safety Relief Valves
BL-74-05	Shipment of an Improperly Shielded Source
BL-74-03a	Failure of Structural or Seismic Support Bolts on Class I Components
BL-74-04	Malfunction of Target Rock Safety Relief Valves
BL-74-03	Failure of Structural or Seismic Support Bolts on Class I Components
BL-74-02a	PWR Main Steam Line Isolation Valves
BL-74-002	Truck Strike Possibility
BL-74-001a	Abnormal Occurrences in Piping Systems
BL-74-001	Valve Deficiencies
BL-73-06	Inadvertent Criticality in a Boiling Water Reactor
BL-73-05	Manufacturing Defects in Boiling Water Reactor Control Rods
BL-73-04	Defective Bergen-Patterson Hydraulic Shock Absorbers
BL-73-03	Defective Hydraulic Shock Suppressors and Restraints

—Cont'd

BL-73-02	Malfunction of Containment Purge Supply Valve Switch
BL-73-01	Faulty Overcurrent Trip Delay Device in Circuit Breakers for Engineered Safety Systems
BL-72-03	Limitorque Valve Operator Failures
BL-72-02	Simultaneous Actuation of a Safety Injection Signal on Both Units of a Dual Unit Facility
BL-72-01	Failed Hangers for Emergency Core Cooling System Suction Header
BL-71-03	Connecticut Yankee Atomic Power Company: Involves Catastrophic Failure of Main Steam Line Relief Valve Headers
BL-71-02	Rochester Gas & Electric Company: Involves PWR Reactor Trip Circuit Breakers
BL-71-01	Niagara Mohawk Power Corporation: Involves Main Steam Isolation Valves

Source: http://www.nrc.gov/reading-rm/doc-collections/gen-comm/bulletins

INDEX

Note: Page numbers followed by "f" indicate figures.

A

Accidental drop, 271
ACI. *See* American Concrete Institute
ACRS. *See* Advisory Committee on Reactor Safeguards
Advanced light-water reactor (ALWR), 42
Advisory Committee on Reactor Safeguards (ACRS), 2
AEC. *See* Atomic Energy Commission
AFW. *See* Auxiliary feed water
Age-related damage, 225–226
Aging mechanism, 217–233
Air handling unit (AHU), 251
AISC. *See* American Institute of Steel Construction
AISI. *See* American Iron and Steel Institute
ALA. *See* American Lifeline Alliance
Allowable stress, 119
ALWR. *See* Advanced light-water reactor
American Concrete Institute (ACI), 10, 18
American Institute of Steel Construction (AISC), 10, 18, 35
American Iron and Steel Institute (AISI), 10, 35, 155–156
American Lifeline Alliance (ALA), 75
American National Standard Institute (ANSI), 16
American Nuclear Society (ANS), 16, 18–19
American Society of Mechanical Engineers (ASME), 1
  ASME B&PV Code, 10–12
  ASME II specifications, 214–215
    material specifications, 215
  ASME III, 12–13
  ASME III D.spec., 169–170
  ASME III pressure boundary, 180
  ASME XI, 16–17
American Society of Testing and Materials (ASTM), 14, 19
American Welding Society (AWS), 16, 19
Anchor system, 126–127
ANI. *See* Authorized Nuclear Inspector
ANII. *See* Authorized Nuclear Inservice Inspector
ANL. *See* Argonne National Laboratory
ANS. *See* American Nuclear Society
ANSI. *See* American National Standard Institute
Argonne National Laboratory (ANL), 83
As-built reconciliation, 131
ASME. *See* American Society of Mechanical Engineers
ASTM. *See* American Society of Testing and Materials
Atomic Energy Commission (AEC), 1–2
Authorized Nuclear Inservice Inspector (ANII), 103–104
Authorized Nuclear Inspector (ANI), 103–104
Auxiliary feed water (AFW), 46–47
Auxiliary system, 220
AWS. *See* American Welding Society

B

Base metal defects, 106
Baseplate, 196
Beyond-design basis event (BDBE), 58
Boiler and Pressure Vessel Code (B&PV Code), 1, 10–12
Braided hose replacement, 244
Break exclusion zones (BEZs), 33
Bulletins, 9

Buried pipe, 73–77
Buried Pipe Integrity Group (BPIG), 76, 247–248

C

Cable trays system, 160–163
  design loads for, 34
  seismic tests, 34
Carbon steel, 215–216
Cast austenitic stainless steel (CASS), 224
Cast-in-place bolts, 183–184
Cavitation, 256
CC. *See* Code cases
CDF. *See* Core damage frequency
CFD. *See* Computational fluid dynamics
CFR. *See* Code of Federal Regulations
CGD. *See* Commercial-grade dedication
Circumferential breaks, 89–90
Climbing, 267–268
Closely spaced bolts, tensile capacity of, 185
Code cases (CC), 10, 239–240
Code of Federal Regulations (CFR), 2–3
Cold functional tests, 112
Commercial-grade dedication (CGD), 214, 259
Computational fluid dynamics (CFD), 49
Concrete
  cracks, 271
  pad, 270
  structures, 171–175
Concrete anchor bolts, 143, 183–184, 195–196
  codes for, 184
  design margin for, 184
  failure of, 184–185
  function of, 184
  parameters, 185
Condenser wall vibration, 240
Conduits, 34, 157–160
Construction completion test, 112
Construction permit (CP), 124
Containment spray (CS), 154–155, 242
Containment structure, 174
Contractor report (CR), 8–9
Control rod drive mechanism (CRDM), 227, 260
Cookbooks, 33, 129
  method, 134
Core damage frequency (CDF), 61–62
Corrosion resistance, 216
CP. *See* Construction permit
CR. *See* Contractor report
Cracked compressor housing, 253
Cracked expansion joints, 255
CRDM. *See* Control rod drive mechanism
CS. *See* Containment spray

D

Damage mechanism, 223–225
  FAC, 230
  neutron embrittlement, 231
Damage-specific inspections, 112
Damp vault wall, 251
Decoupling, 126–127
Deterministic analysis, 53–58
Dissimilar metal welds (DMW), 226
Distribution system. *See* Suspended system
DMW. *See* Dissimilar metal welds
Door window, 268
Double-seal design, 253
Dynamic loads, 29–32, 45–63. *See also* Load; Static loads
  fluid transients, 45–53
  seismic design, 53–63

E

East–west (EW), 208
ECCS. *See* Emergency core cooling system
EDG. *See* Emergency diesel generator
Electric Power Research Institute (EPRI), 235
Emergency core cooling system (ECCS), 89
Emergency diesel generator (EDG), 30–31, 219
Environmental qualification (EQ), 220
EPRI. *See* Electric Power Research Institute
EQ. *See* Environmental qualification

EW. *See* East–west
Examination, 103–112

F

FA. *See* Functionality Assessment
FAC. *See* Flow-accelerated corrosion
Failure mode and effects analysis (FMEA), 274–275
Fall arrestor, 268
Fall protection system, 267
Fast Fourier transform (FFT), 64
Fatigue, 77–85
  cracking, 31
  failure, 77
FEA. *See* Finite element analysis
Feedwater transient, 276–277
FFS. *See* Fitness-for-service
FFT. *See* Fast Fourier transform
Final safety analysis report (FSAR), 4, 124, 273
Finite element analysis (FEA), 70, 127–128, 166–167, 264–265, 264f
Fire protection system, 220
FIRS. *See* Foundation input response spectra
Fitness-for-service (FFS), 235, 237
FIV. *See* Flow-induced vibration
Flange joints, 147
Flaw-tolerant design, 83–84
FLEX Coping Strategies. *See* Flexible Coping Strategies
Flexible Coping Strategies (FLEX Coping Strategies), 60–61
Flow diagrams. *See* Process and Instrumentation Diagrams (P&IDs)
Flow-accelerated corrosion (FAC), 111, 217, 230
Flow-induced vibration (FIV), 27, 31, 64, 239
Fluid transients, 45–53
FMEA. *See* Failure mode and effects analysis
Foundation input response spectra (FIRS), 59
Fragility, 62
FSAR. *See* Final safety analysis report
Functional capability limit, 138
Functionality Assessment (FA), 20–21

G

GALL. *See* Generic aging lessons learned
GDC. *See* General design criteria
General design criteria (GDC), 3–4, 54
Generic aging lessons learned (GALL), 218
Generic Implementation Procedure (GIP), 34
Generic letter (GL), 9, 214, 270
GIP. *See* Generic Implementation Procedure
GL. *See* Generic letter
GMRS. *See* Ground motion response spectra
Grating, 168
  temporary load on, 198–200
Ground motion response spectra (GMRS), 59
Guillotine breaks. *See* Circumferential breaks

H

Hanger, 241. *See also* Snubber
Hatch cover, 190f
  tornado pressure on, 189–190
HAZ. *See* Heat-affected zone
HCLPF. *See* High confidence of low probability of failure
HE line break loads. *See* High-energy line break loads
Heat, Ventilation, and Air-Condition (HVAC), 28, 35, 154–157
  design, 35
  loads, 35
  seismically qualification, 35–36
Heat-affected zone (HAZ), 108
Heavy lift, 168
High confidence of low probability of failure (HCLPF), 62
High-energy line break loads (HE line break loads), 31
High-pressure coolant injection (HPCI), 69

High-strength bolt, 183
Honeycomb whip restraints, 93
Hot concrete penetration, 266
Hot functional test, 112
Housekeeping, 249
HPCI. *See* High-pressure coolant injection
HVAC. *See* Heat, Ventilation, and Air-Condition
Hydraulic transient loads, 30
Hydrostatic test, 114–117
  pressure, 32
Hydrotest. *See* Hydrostatic test

I

ID. *See* Inner diameter
IEEE. *See* Institute of Electrical and Electronics Engineers
IER. *See* Industry Event Report
IN. *See* Information Notices
In-service bolt replacement, 259–260
In-service inspection (ISI), 16–17, 33, 239
In-service surveillance testing (IST), 17, 69–70
In-structure response spectra (ISRS), 29–30, 131–132
Industry Event Report (IER), 273
Industry operating experience (IOE), 235, 272
  affecting plants, 273
  departments using, 273
  examples, 275–282
  FMEA, 274–275
  guidance for performing, 275
  industry-wide effort, 273
  maintenance, 274
  MR mandatory, 274
  plants implementing FMEA, 275
  plants implementing MR, 274
Information Notices (IN), 9
Initial core loading test, 113
Inner diameter (ID), 101
INPO. *See* Institute of Nuclear Power Operations
Inquisitive attitude, 265
Inservice Testing. *See* In-service surveillance testing (IST)
Institute of Electrical and Electronics Engineers (IEEE), 16, 19–20
Institute of Nuclear Power Operations (INPO), 235
Internal missile strike, 31
International Society of Automation (ISA), 151
IOE. *See* Industry operating experience
ISA. *See* International Society of Automation
ISI. *See* In-service inspection
ISRS. *See* In-structure response spectra
IST. *See* In-service surveillance testing

J

Japan Atomic Energy Research Institute (JAERI), 95
Jet and whip restraints, 165
Junction boxes, 157, 159–160

L

Large early-release frequency (LERF), 61–62
Leak before break (LBB), 27, 97–102
Leak testing, 112–122
Leakage crack, 101
Leaking
  expansion joint, 245–246
  gasket, 254
  threads, 254
LERF. *See* Large early-release frequency
Licensee Event Report (LER), 277
Lifting
  devices, 168–169
  pads, 266
Lighter-weight commodities, 144
Limited inventory break, 93
Line pipe, 148
Liquid penetrant testing (Liquid PT), 104
Load
  combinations, 39–45
  suspended systems, 27–36
  system supports, 36–39
Long-term scaffold service, 204–205

Loss of coolant accident (LOCA), 35
Low alloy steel, 213, 215–216, 221–222

M

Magnetic particle testing (MT), 104
Main feed water (MFW), 46–47
Main steam isolation valve (MSIV), 37, 265
Maintenance Rule (MR), 273
Material Reliability Program (MRP), 87–88, 282
Mechanically induced vibration, 63–64
MFW. *See* Main feed water
Missiles, 30–31
Modal analysis, 131–132
Moisture damage, 238

N

National Electrical Manufacturers Association (NEMA), 159–160
NDE. *See* Nondestructive examinations
Near-Term Task Force (NTTF), 59
NEI. *See* Nuclear Energy Institute
NEMA. *See* National Electrical Manufacturers Association
Neutron
  damage steel, 231
  embrittlement, 231–232
New single-point vulnerability, 275–276
NNS. *See* Non-Nuclear Safety
Nominal pipe size (NPS), 108
Non-Nuclear Safety (NNS), 22
Nondestructive examinations (NDE), 103–104
Nondestructive testing (NDT). *See* Nondestructive examinations (NDE)
Nonmetallic joint, 245
Nonmetallic materials, 216–217
Nonsafety systems, 24
Nonsafety-related welds, 181
North–south (NS), 208
NPS. *See* Nominal pipe size
NQA-1 program. *See* Nuclear QA program
NRC. *See* Nuclear Regulatory Commission
NS. *See* North–south
NTTF. *See* Near-Term Task Force
Nuclear Energy Institute (NEI), 235
Nuclear power plants, 213
  aging mechanism, 217–233
  bolting
    to concrete, 183–186
    to steel, 182–183
  case study, 207–209
  component clearances, 209–212
  equipment lift, 188–189
  finite element structural analyses, 187–188
  fire protection pipe, 211f
  materials in, 213–217
  pipe whip, 196–198
  scaffolding, 200–207, 203f
  structural steel, 165–170
  structural welding, 175–182
  supporting concrete, 171–175
  suspended systems in, 27–28
  temporary load on grating, 198–200
  tornado pressure on hatch cover, 189–190
  welded steel support structure, 190–191
Nuclear QA program (NQA-1 program), 109
Nuclear Regulatory Commission (NRC), 1–2, 128, 235–236
NUREGs, 8–9

O

O&M Code. *See* Operation and Maintenance Code
OBE. *See* Operating Basis Earthquakes; Operational basis earthquake
OD. *See* Operability determination; Outer diameter
Operability, 235–238
  assessment, 235
Operability determination (OD), 20–21, 235

Operating Basis Earthquakes (OBE), 42–43
Operation and Maintenance Code (O&M Code), 10, 235–236
  O&M 3 criteria, 66–67
Operational basis earthquake (OBE), 29–30, 55, 133
OSHA, 167–168, 201
Out-of-calibration torque wrench, 252
Out-of-tolerance snubber, 240–241
Outer diameter (OD), 101
Overlap method, 127
Overpressure, 242
Overtorqued bolt, 252–253

P

P&IDs. *See* Process and Instrumentation Diagrams
Periodic leak tests, 113
Pinhole leak, 277
Pipe bending, 147
Pipe rupture, 88–96
Pipe rupture hazards analysis (PRHA), 88
Pipe stress analysis, 125–126
Pipe uplift, 247
Pipe whip, 196–198
Pipe whip restraint (PWR), 47
Piping system, 28, 238–248
  allowable stresses, 138–139
  analyzation, 129
  as-built reconciliation, 131
  ASME and AISC, 125
  boundary, 125
  concrete anchor bolts, 143
  cookbook method, 134
  design codes, 124–125
  engineering of, 123
  equipment nozzles, 126–127
  fabrication and construction requirements, 146–147
  FEA software, 129
  flange joints, 147
  functions, 124
  high operating temperatures, 137–138
  instrument, 124
  key attributes, 130
  lighter-weight commodities, 144
  load-rated supports, 142
  loads and load combinations, 136–137
  modal analysis, 131–132
  nuclear power industry, 139
  overlap method, 127
  pipe
    bending, 147
    piping models, 130
    stress analysis, 125–126
    supports, 145–146
  postinstalled anchors, 144–145
  protocols for, 128
  qualification requirements, 138
  seismic analysis, 131–133
  snubbers, 140–142
  software analysis, 127–128
  standard span method, 133–134
  standard supports, 139–140
  stress analysis, 127, 143
  surface appearance, 147–148
  terminal anchors, 146
  V&V, 128–129
PMDA. *See* Proactive materials degradation assessment
Pneumatic test, 118
Post-water hammer actions, 243
Postaccident debris accumulation, 270–271
Power ascension test, 113
Pre-staged bolt removal, 265
Preoperational vibration test, 66
Pressure uprate, 258
Pressurized water reactor (PWR), 123, 171, 215–216, 241
PRHA. *See* Pipe rupture hazards analysis
Primary water stress corrosion cracking (PWSCC), 215–216, 226–227, 263, 280–281
  causes, 227–228
  components, 227
  issue, 229–230
Proactive materials degradation assessment (PMDA), 232
Probabilistic analysis, 58–63

Process and Instrumentation Diagrams (P&IDs), 24
Prying action, 185
Pump vane passing frequency, 70
PWR. *See* Pipe whip restraint; Pressurized water reactor
PWSCC. *See* Primary water stress corrosion cracking

Q

Qualification of Mechanical Equipment (QME), 10
Quality assurance, 213
Quality control examinations (QC examinations), 103
Quantified leakage test, 122

R

Raceways, 158
Rack seismic interaction, 268–269
Radial arm hoist assemblies (RAHAs), 260, 261f
Radiation resistance, 216
Radiographic testing (RT), 104
RAHAs. *See* Radial arm hoist assemblies
RCRBSJ. *See* Research council on riveted and bolted structural joints
RCS. *See* Reactor–coolant system
Reactor pressure vessel (RPV), 263
Reactor–coolant system (RCS), 23, 278–280
Regulations, Codes, and Standards, 10–21
  classification, 21–26
  requirements, 1–21
Regulatory Guide (RG), 4–5, 7–8, 54
Research council on riveted and bolted structural joints (RCRBSJ), 183
Residual heat removal (RHR), 31–32, 278–280
Resonance
  acoustic, 64–65
  structural, 65
RG. *See* Regulatory Guide
RHR. *See* Residual heat removal
RPV. *See* Reactor pressure vessel
RT. *See* Radiographic testing

S

Safe shutdown earthquake (SSE), 29–30, 133, 167
Safety analysis report (SAR), 1
Safety class
  ducting examination, 112
  Safety Class 1, 22
  Safety Class 2, 22–23
  Safety Class 3, 22–23
Safety-related
  SSCs, 10
  welds, 181
SAM. *See* Seismic anchor movement
SAR. *See* Safety analysis report
Scaffolding, 200–207, 203f
  configuration study, 206f
  in nuclear power plant, 203f
Scanning electron microscope (SEM), 106
SCC. *See* Stress corrosion cracking
SCs. *See* Seismic categories
Seismic analysis, 131, 157
Seismic anchor movement (SAM), 29
Seismic categories (SCs), 25
  SC-I structures, 32
Seismic design, 53–63
  deterministic analysis, 53–58
  probabilistic analysis, 58–63
Seismic interactions, 167–168
Seismic margins assessment (SMA), 59
Seismic probabilistic risk assessment (SPRA), 58–59
Seismic Qualification Utilities Group (SQUG), 34
SEM. *See* Scanning electron microscope
Sensitivity, 114
Service Levels, 22
SG. *See* Steam generator
SIF. *See* Stress intensification factor
Significant Operating Experience Report (SOER), 273
SMA. *See* Seismic margins assessment
Snubbers, 140–142, 240–241
Snug-tight joint, 182–183
Socket weld, FIV in, 239
SOER. *See* Significant Operating Experience Report

Soft foot, 63–64
Software analysis, 127–128
Soil–structure interaction model (SSI model), 55
Special analysis
 fatigue, 77–85
 LBB, 97–102
 pipe rupture, 88–96
 thermal stratification, 86–88
SPRA. *See* Seismic probabilistic risk assessment
Square root sum of squares (SRSS), 42
SQUG. *See* Seismic Qualification Utilities Group
SRP. *See* Standard Review Plan
SRP-LR applications. *See* Standard review plan for review of license renewal applications
SRSS. *See* Square root sum of squares
SS. *See* Stainless-steel
SSC. *See* Structures, systems, and components
SSE. *See* Safe shutdown earthquake
SSI model. *See* Soil–structure interaction model
Stainless-steel (SS), 213, 215–217, 245
Standard Review Plan (SRP), 4, 29
Standard review plan for review of license renewal applications (SRP-LR applications), 218
Standard span method, 133–134
Static loads, 28–29
Static seismic analysis method, 132–133
Steam and power conversion system, 220
Steam generator (SG), 46–47, 105, 227, 241
Steel bolting, 182–183
Steel plates, embedded, 173–174
Steel structure, 165–170
 bolting to steel, 182–183
 structural welding, 175–182
Stress analysis subsystem, 127
Stress corrosion cracking (SCC), 105, 112, 217
Stress intensification factor (SIF), 67
Structural capability pressure test, 122
Structures, systems, and components (SSC), 1, 27–28, 124, 167–168, 235
 mechanical equipment, 252–265
 operability determination, 235–236
 piping system, 238–248
 raceways and ducts, 248–252
 structure, 265–272
 tubing system, 238–248
Subcompartment pressurization, 204, 208
Surface examinations, 104, 108–109
Suspended systems, 27–36, 123
 cable trays system, 160–163
 conduit system, 157–160
 HVAC system, 154–157
 piping system, 123–148
 tubing system, 148–154

T

Tack welding, 181
Technical specification (TS), 236
10 CFR, 2–3
Tensile capacity, 185
Testing, 22
Thermal sleeves, 85, 280–281, 280f
Thermal stratification, 86–88
Tornado load, 30–31
Tornado missile, 263–265
Tornado pressure, 189–190
Tracking system, 127
Trapped air pockets, 242
TS. *See* Technical specification
Tsunami, 60
Tubing slope, 151
Tubing system, 28, 148–154, 238–248. *See also* Piping system

U

U-bolt, 191–192
Ultrasonic examination (UT examination), 104–105
Unresolved Safety Issue (USI), 45
Upper-level safety requirements, 3–4
USI. *See* Unresolved Safety Issue
UT examination. *See* Ultrasonic examination

V

Valve cavitation, 257
Valve seat failure, 277–278
Ventilation system, 220
Verification and validation (V&V), 128–129
Vibrating fan, 268
Vibration
 monitoring program, 261–262
 testing and monitoring, 63–73
Vibration monitoring groups (VMGs), 69
Visual testing (VT), 104
VMGs. *See* Vibration monitoring groups
Volumetric examinations, 104–105

W

Water hammer, 238–239
Weld seam strength test, 122
Welded steel support structure, 190–191
Welding, 175–182
Welding Research Council (WRC), 74
Whip restraints, 91–93

Y

Yoke-to-Bonnet stud failure, 276

Z

Zone of influence (ZOI), 91, 93

FSC
www.fsc.org
MIX
Papier | Fördert
gute Waldnutzung
FSC® C083411

Kolibri 360 GmbH
Ferdinand-Jühlke-Straße 7
99095 Erfurt, Deutschland
produktsicherheit@kolibri360.de